Cambridge Elements

Elements in Philosophy of Science
edited by
Jacob Stegenga
University of Cambridge

SCIENTIFIC REALISM

Timothy D. Lyons
Indiana University Indianapolis

Shaftesbury Road, Cambridge CB2 8EA, United Kingdom

One Liberty Plaza, 20th Floor, New York, NY 10006, USA

477 Williamstown Road, Port Melbourne, VIC 3207, Australia

314–321, 3rd Floor, Plot 3, Splendor Forum, Jasola District Centre,
New Delhi – 110025, India

103 Penang Road, #05–06/07, Visioncrest Commercial, Singapore 238467

Cambridge University Press is part of Cambridge University Press & Assessment,
a department of the University of Cambridge.

We share the University's mission to contribute to society through the pursuit of
education, learning and research at the highest international levels of excellence.

www.cambridge.org
Information on this title: www.cambridge.org/9781009507226

DOI: 10.1017/9781108588430

© Timothy D. Lyons 2025

This publication is in copyright. Subject to statutory exception and to the provisions
of relevant collective licensing agreements, no reproduction of any part may take
place without the written permission of Cambridge University Press & Assessment.

When citing this work, please include a reference to the DOI 10.1017/9781108588430

First published 2025

A catalogue record for this publication is available from the British Library

ISBN 978-1-009-50722-6 Hardback
ISBN 978-1-108-70665-0 Paperback
ISSN 2517-7273 (online)
ISSN 2517-7265 (print)

Cambridge University Press & Assessment has no responsibility for the persistence
or accuracy ofURLs for external or third-party internet websites referred to in this
publication and does not guarantee that any content on such websites is, or will
remain, accurate or appropriate.

Scientific Realism

Elements in Philosophy of Science

DOI: 10.1017/9781108588430
First published online: January 2025

Timothy D. Lyons
Indiana University Indianapolis
Author for correspondence: Timothy D. Lyons, tdlyons@iu.edu

Abstract: The scientific realism debate directly addresses the relation between human thought and the reality in which it finds itself. A core question: Can we justifiably believe that science accurately describes the reality that lies beneath the limits of human experience? Exploring this question, this Element begins at the most foundational level of scientific realism, the endeavor to justify belief in the existence of unobservables by way of abduction. Raising anti-realist challenges, some much discussed in the literature but also some generally overlooked, it works its way toward more refined variants of scientific realism. Because scientific realism is the default position of many – scientific realists themselves often assuming it is the default position of scientists – the emphasis will be on the challenges. Those challenges will also motivate the variants of scientific realism traced. The Element concludes with a brief articulation of the author's own position, Socratic scientific realism.

Keywords: no-miracles argument, pessimistic meta-induction, the argument from the bad lot, theoretical virtues, Socratic scientific realism

© Timothy D. Lyons 2025

ISBNs: 9781009507226 (HB), 9781108706650 (PB), 9781108588430 (OC)
ISSNs: 2517-7273 (online), 2517-7265 (print)

Contents

1 Believing Existence Abductions   1

2 Believing the Best Explanation: The Realist's Move
  to Comparative Inference   10

3 Supraempirical Virtues and Their Prospects for Justifiably
  Excluding Competitors   25

4 Truth and the Argument from the Bad Lot   40

5 The Realist Justification for Epistemic Privilege: The
  No-Miracles Argument   48

6 Conclusion and Epilogue: Socratic Scientific Realism   65

   References   71

1 Believing Existence Abductions

1.1 Abduction and the Existence of Unobservables

The central claim of scientific realism is that science endeavors to accurately describe reality beyond the realm of what we have observed or even can observe, a reality that is not dependent on our beliefs about it. Another vital assertion is a methodological claim about how this is done. Scientific realists claim that, to "get at" an unobservable reality that does not depend on our beliefs, science deploys a particular kind of inference, abduction. Most crucially, for the scientific realist, this mode of inference is essential to justifying our beliefs about that unobservable reality.

In the current section, I will launch our inquiry by discussing the most foundational form of scientific realism and its favored inference, abduction. I will use this foundational form of realism to provide at least a preliminary glimpse of four challenges to scientific realism. We will later see these challenges prompting realists toward more sophisticated variants of their position – for instance from realism concerning existence, to a realism concerning theories; from a realism that explains physical phenomena, to a realism that explains the *success* of theories, and so on. In the course of unfolding such variants of scientific realism, I will provide more thorough articulations of the first four challenges, those that are introduced in Sections 1.2–1.5, and I will introduce four more challenges to scientific realism as they become relevant. In addition to tracing increasingly sophisticated variants of realism and along the way raising eight specifiable challenges to scientific realism, I will emphasize an important feature of these challenges: these challenges can be dovetailed in significant ways, ways that pose increasingly strong threats to the variants of scientific realism I will be tracing. With that admittedly brief overview of where we are heading, I begin by introducing that mode of inference our realist takes to be fundamental, that which is meant to justify our beliefs about unobservable reality as science describes it: abduction.

Abduction is taken to be a form of *explanatory* reasoning. Although pointed to by William Whewell – and possibly even Aristotle – abduction was given its name, famously articulated, and advocated by C.S. Peirce (1958). While Peirce construes it in several distinct ways, the most significant articulation, or at least the one that has caught on, is as follows: We begin with a "surprising" observation, (O). A state of affairs is postulated, and that postulate, (P), would render (O) "a matter of course." We conclude that "we have reason to suspect" that (P) obtains (1958, p. 189). Embracing this form of inference, the scientific realist will (eventually) lift the abductive conclusion an epistemic step higher than Peirce, shifting up from "we have a reason to *suspect*" P to the claim that "we are justified in believing" it. This kind of inference is meant to take us to conclusions about the world that go beyond and beneath the objects we access by way of our immediate sense experience. Moreover, as Michael Devitt, a contemporary scientific realist, sees it, "in the discussion of unobservables – the debate about *scientific realism* – the main controversy has been over *existence*" (2013, p. 257) [the second italics are mine].

Accordingly, for now, we can see abduction as an explanatory inference to an *existential* claim, in particular. Plugging in a scientific example: A surprising fact, O, say the photoelectric effect, is observed, and a postulate, P, about the unobservable, U, say, that light is made up of photons, is put forward. The existence of what is posited, U, photons, would make the observed phenomena, O, the photoelectric effect, "a matter of course." In this case, we have justification for believing that the unobservable entity U, the photon, exists. In short, for the realist, abduction is that mode of inference used in science to get at the reality that is not dependent on our beliefs and underlies human sense experience.

During most of the twentieth century, a primary concern of philosophers was to identify the mode of inference that distinguishes science from other areas of inquiry, a method of inference that can stand as a demarcation criterion between science and non-science (e.g., Reichenbach, 1930; Popper, 1959; Kuhn, 1974; Lakatos, 1970, 1974). Intriguingly, however, the scientific realist takes us in the opposite direction: the scientific realist considers it to be a, if not *the*, significant virtue of abduction, and a virtue of their position in general, that this same mode of inference is also heavily employed outside of science. That is, although abduction is employed to justify belief in the existence of potentially exotic unobservables posited in science, abduction is not taken by scientific realists to constitute some kind of esoteric mode of reasoning, one employed only by scientists, or even only scientists and philosophers. On the contrary, abduction is construed, not only as an extension of our best mode of ampliative inference, but as a form of reasoning that plays a pivotal role in our everyday lives.

I dare say this claim about our everyday dependence on abduction is the feature that realists take to be its primary, if not sole, justification. Bas van Fraassen, though an anti-realist, offers a well-known illustration of such everyday reasoning: "I hear scratching in the wall, the patter of little feet at midnight, my cheese disappears – and I infer that a mouse has come to live with me. Not merely that these apparent signs of mousely presence will continue, not merely that all the observable phenomena will be as if there is a mouse; but that there really is a mouse" (1980, pp. 19–20). The existence of the mouse makes those phenomena – the scratching, the sound of the patter, the disappearing cheese – a matter of course: the existence of the mouse explains the phenomena. Likewise, says the realist, in science, it is that same mode of inference that takes us to the existence of unobservables. As Devitt expresses it: "by supposing that the unobservables of science exist, we can give good explanations of the behavior and characteristics of observed entities, behavior and characteristics which would otherwise remain inexplicable" (2013, p. 261). Devitt calls this argument for scientific realism the "basic abductive argument" (2010, p. 78). This "basic argument uses realism" in this case, the existence of unobservables, "to explain observed phenomena" (2010, p. 78).

This abductive justification for belief in the existence of unobservables posited in scientific theories has a long-standing history among scientific realists. Indicating something along these lines, in 1962, Wilfred Sellars wrote, "to have good reasons for espousing a theory which postulates the existence of unobservable entities is to have good reason for saying that these entities really exist" (1962, [2017] 98). More specifically, with an explicit reference to Peirce, J.J.C. Smart wrote in 1963, "if we interpret a theory in a realist way, then we have no need for ... a cosmic coincidence: it is not surprising that galvanometers and cloud chambers behave in the sort of way they do, for if there really are electrons, etc., this is just what we should expect. A lot of surprising facts no longer seem surprising" (1963, p. 39). Similarly in 1968, Smart wrote, "If there were no such things," that is, if it were not the case that "there really are electrons or whatever is postulated by the theory," we "would have to suppose that there were innumerable lucky accidents about the behavior mentioned in the [theory's] observational vocabulary, so that [the phenomena] behaved miraculously *as if* they were brought about by the non-existent things ostensibly talked about in the theoretical vocabulary" [italics in original] (1968, pp. 150–1). He later wrote, "the scientific realist looks at the facts on the observational level and asks why they are as they are" (1979, pp. 364–5); the realist "is not satisfied with an accidental observation level" (p. 365). He says, "only if there is a realistic theory can we avoid supposing an implausible cosmic coincidence on the observational level" (p. 373). Devitt gives a cut-to-the-chase example that accords with this: "why are all the observations we make just the sort we would make if there were atoms? Answer: because there *are* atoms" (Devitt, 2010, p. 73).

Scientific realists tend to take it for granted that scientists are themselves scientific realists: as Hilary Putnam puts it, "Realism is, so to speak, 'science's philosophy of science'" (1976, p. 193). Smart likewise indicates this, appealing to the reasoning of scientists themselves: "by postulating unobservable particles, and so on, and by stating a relatively small number of laws pertaining to these, a scientist can explain the untidy and multifarious facts about the macro-level in a relatively simple and unified manner" (1979, p. 365). Likewise, Wesley Salmon, wanting an "empirical approach to the philosophical problem" of whether unobservables exist, focuses on "the evidence and arguments that convinced scientists of the reality of unobservable entities" (1984, pp. 213–4). Ernan McMullin writes that the "best case for scientific realism" (1984, p. 26) is as follows: "Scientists construct theories which explain the observed features of the physical world by postulating models of the hidden structure of the entities being studied. This structure is taken to account causally for the observable phenomena" (p. 27).

An important feature here is that these realists are denying that the inference takes place at a meta-scientific, philosophical, level. McMullin, making this explicit, writes "the argument is properly carried on at one level only, the level of the scientist" (1991, p. 104). These realists take this to be a positive feature of this basic realist inference over other variants we will see later. Concurring with McMullin, Devitt notes that an "advantage of" this "basic argument" for realism "is that it makes clear that... the use of abduction to justify realism is not at some 'philosophical' level above science" (2010, p. 78, ftnt 23). For contrast with another version of realism to be discussed later, it is important to note that the phenomena calling for explanation are those of the natural world, and that which does the explaining are the unobservable entities in a scientific theory: what the latter explain are natural phenomena. On this view, we can simply take scientific theories and the entities posited therein at "face value," a phrase Putnam uses (1976, p. 193). He writes that the realist "argues that science should be taken at 'face value' ... and that science taken at 'face value' *implies* realism" (p. 193).

One more feature of scientific realism is that it is nearly always touted as an empirically testable position. And in accord with Salmon's desire for an "empirical approach," scientific realism, even at this level, is taken to be testable. Devitt writes that in this basic argument, "the explanation of observed phenomena, like any explanation, is *tested* by its observational success. So according to the basic argument, realism," belief in the existence of unobservables arrived at by abduction, "*is* successful..." (Devitt, 2010, 78) [original italics]. (On the empirical testability of, and attempts to test scientific realism, see also Putnam, 1976; Laudan, 1981; Lyons 2002, 2006b; Haufe, 2016; Lyons and Vickers, 2021.) I will now introduce four challenges to scientific realism as thus far construed.

1.2 Challenge 1: van Fraassen against the Demand for Explanation

As noted, Bas van Fraassen is a scientific antirealist. Among the numerous challenges he offers against scientific realism, one that has received rather minimal attention compared to others he has put forward is his *argument against the demand for explanation*. This argument is interspersed throughout the first part of his framework-setting book, *The Scientific Image* (1980). He writes that explanatory "arguments for realism succeed only if the demand for explanation is supreme – if the task of science is unfinished, *ipso facto,* as long as any pervasive regularity is left unexplained" (1980, p. 23). Van Fraassen paraphrases the arguments we have seen from Smart as follows: "The regularities in the observable phenomena must be explained in terms of deeper structure, for otherwise we are left with a belief in lucky accidents and coincidences on a cosmic scale" (p. 25). Against this he suggests "that if the demand for explanation implicit" in these realist abductions "were precisely formulated, it would at once lead to absurdity" (p. 25). Consider Smart's comment we saw earlier, that the realist "is not satisfied with an accidental observation level" (1979, p. 365). One can take van Fraassen's point to be that, on pain of inconsistency, the realist, going one level deeper, so to speak, would have to be dissatisfied with a new "accidental" correlation. Going deeper yet to resolve *that* "lucky accident," avoiding "an implausible cosmic coincidence" at that new "level," the realist would be compelled to go deeper yet to a third level. And this continues, suggests van Fraassen, ad infinitum. If, instead, the realist counters this infinite slide by halting the demand for explanation, at, say, that third or any other deep level, the decision to do so would be arbitrary. Moreover, that arbitrary blocking of the explanatory regress would do nothing to solve the realist's original problem of facing an unexplained regularity. However, says van Fraassen, we need not choose between an infinite regress and an arbitrary stopping point; for that dichotomy is the result of the realist demand for explanation. Rather, instead, we can reject that demand and – a further point that can be gleaned from van Fraassen's comments is that – we can opt instead for the *epistemically non-arbitrary belief* regarding co-relations or regularities at the observable level.

Of course, scientists will posit what they posit, and with good reason; van Fraassen isn't trying to change the way science is practiced. In fact, van Fraassen is happy to point out, in his favor, that there are scientists who do not follow the realist's demand for explanation. He writes, for instance, "such an unlimited demand for explanation leads to a demand for hidden variables which runs contrary to at least one major school of thought in twentieth-century physics" (1980, p. 23). But scientific practice aside, for the moment, and in terms of belief in particular, with which the realist is concerned: the threat of the explanatory regress can be halted by halting belief at what the theory tells us about that to which we humans have the best epistemic access, observables.

1.3 Challenge 2, a Glimpse: The Historical Argument against Realism

In 1859, J.S. Mill anticipated a variant of the historical threat against scientific realism, what is taken to be a pessimistic meta-induction. He wrote, "every age having held many opinions which subsequent ages have deemed not only false but absurd . . . it is as certain that many opinions, now general, will be rejected by future ages, as it is that many, once general, are rejected by the present" (1859 [1998, p. 23]). In 1895, noting that it is "difficult . . . to-day to understand the state of mind of learned men" in the past, Leo Tolstoy wrote that, unless

> our century forms an exception (which is a supposition we have no right to make), it needs no great boldness to conclude by analogy [or "to foresee"] that among the kinds of knowledge occupying the attention of our learned men and called science, there must necessarily be some which will be regarded by our descendants much as we now regard the rhetoric of the ancients and the scholasticism of the Middle Ages. (1895, [1903, p. 105])[1]

Mindful of the basic abduction and expressing it in terms of unobservables, specifically, Devitt offers an articulation of "a basic version" of this historical induction: "the unobservables posited by past theories do not exist; so, probably, the unobservables posited by present theories do not exist" (2013, p. 86). Phlogiston, which was taken to be the stuff of flames; caloric, taken to be the substance of heat; and the luminiferous ether, taken to be the medium through which light travels – these are typical candidates discussed in the context of this historical argument. Having seen previously that realists tout their position as empirically testable, the historical argument can be understood as taking that claim seriously. The idea is that just as these unobservable entities factored into abductions in the past but do not exist, we have inductive grounds to infer that, so too, the unobservable entities factoring into our contemporary abductions do not, or are unlikely, to exist. Larry Laudan – the other anti-realist, who, with van Fraassen, brought the contemporary framework to the realism debate – provides what has come to be an infamous list of such entities in his (1981). I will discuss this historical argument in more detail in Section 4.3. I mention it here, first, because, in contrast with van Fraassen's argument against the realist's explanatory demand, it is far more widely discussed in the literature and is undoubtably one of the two primary threats to scientific realism, to which we will return.

---

[1] To my knowledge Stathis Psillos (2018) is the first to bring Tolstoy's (1895 [1903]) comments into the contemporary debate. While Psillos cites the (1895 [1903]) text, translated from French, I've inserted in brackets "foresee," which is the key term in Leo Wiener's 1905 translation of Tolstoy's text "from the Original Russian" (1893 [1905, p. 45]). The use of "foresee" makes clear that Tolstoy's argument is, like Mill's, an induction. The structure of the historical argument will become important in later sections.

Second, the historical argument gives us further grounds for refraining from belief beyond the observable. Third, even the brief expression of the aforementioned examples indicates a point I will now explore: with the positing of unobservable entities come descriptions of those entities.

1.4 Challenge 3, a Glimpse: The Threat of a Low Degree of Implication and a Shift to "T"

Abduction as described in Section 1.1 too easily invites the impression that the claim to justifiably believe that U exists is extremely minimal. Yet, as just indicated with phlogiston, caloric, and ether, that minimality is illusory. First, of course, one would believe the postulate, P, which is at the very least, in the cases of concern, a general existential statement: "U exists." However, second, conjoined to P would be descriptions of properties attributed to U. These would include simple properties, say, that the photon is a light packet, but also numerous others, such as photons have no mass, and so on; and, considering contemporary science, especially, what would be said of photons would be massively complicated by quantum theory.

Toward making this salient, we can note first that, in the realist explanatory argument, that which does the explaining is that which the realist claims we can justifiably believe. Second, we can motivate a sophistication in our characterization of realism by introducing a third challenge. Although best known for its historical argument against the scientific realists, Laudan's (1981) article also included a separate argument, one against the realist's "downward path" (p. 29). To understand Laudan's phrase here, note that the "path" is an argumentative one; and more specifically, for Laudan, explanation requires a deduction. So we are to visualize a deductive argument that begins, at "the top," so to speak, with posits regarding unobservables; from there, following the path *downward* through the deduction, we would arrive at "the bottom," that is, the conclusion of the deduction, where that conclusion contains statements describing observables. This is how one can conceptualize Laudan's dubbing the realists' *explanatory* task as a downward path: as a deduction going from posits regarding unobservables, which do the explaining, "downward" argumentatively, to a conclusion about observables, that is, to that which is being explained. And a key takeaway in Laudan's critique of the realists' downward path, paraphrased for present purposes, is that merely positing that U exists (or that a term genuinely *refers*) does nothing to (deductively) guarantee that we will be led to any statements about observable phenomena.[2] Note that, although Laudan is

---

[2] Laudan argues that reference is insufficient for *success*. My current focus is the need for U to bring about the *observed phenomena*. The difference between the latter and "success" will become important in Section 5.1 and after.

demanding a relation of deductive entailment between what is doing the explaining and what is explained, in Section 1.1, we characterized abduction using the looser Peircean notion of abduction in which P need only render observed phenomena "a matter of course"; and we might charitably interpret this, on the realist's behalf, to be less demanding than entailment. Even then, however, we appear compelled to take it to mean that P will at least *render likely* that which it is explaining. This bears significantly on our realist's explanatory demand. In terms I've used elsewhere (Lyons, 2003), the *degree of implication* possessed by U alone for the phenomena – that is, the *degree* to which the mere posit "U exists" *implies* the phenomena – is so low it cannot render the phenomena likely. Our realist would lack the downward argumentative path – from "U" to the phenomena – needed for U to explain those phenomena. In order to solve this third explanatory challenge, that of needing to claim U would make the phenomena likely – and mindful that the realist's explanatory argument claims justified belief for what is doing the explaining – our realist must claim justified belief for more than just U, adding at least a set of property descriptions about U, as in the previous paragraph. Let's dub this minimal set of property descriptions "a theory," T. Although this explanatory challenge forces us to broaden the scientific realist's commitment from U to T (which includes U), Putnam's "face value" (1976, p. 193) realism endures. Embracing face value realism now with regard to T, Michael Levin says scientific realism should be taken to amount to "simply believing scientific theories" (1984, 134). Since it would follow that the entities posited therein exist, despite the extension from believing "U exists" to believing a theory about U, the core of realism regarding unobservables is retained. For now, we will simply let "T" suffice as the new realist commitment and revisit the issue of the downward path where it again becomes relevant in a later section.

1.5 Challenge 4, a Glimpse: The Competitor Thesis and Underdetermination

The solution we've just noted to Challenge 3 opens the way for a fourth challenge I will call "The Competitor Thesis," which itself gives rise to what we can call issues of underdetermination of theories by data. Along with the historical argument indicated in Section 1.3, the argument from underdetermination constitutes the other primary argument against scientific realism. We have now been forced to shift from a realist claim that we can justifiably believe a kind of unobservable exists to the claim that we can justifiably believe a theory, T. Take the notion of a *distinct alternative* to T to be a theory such that it contradicts T or more strongly would, if it genuinely describes reality,

Scientific Realism   9

render *T* patently false. While any number of distinct alternatives would fail empirically, take the term *competitor* to denote a distinct alternative that can replace *T* in an abduction, that can, no less than T, make the relevant phenomena "a matter of course." To remain neutral, for the moment, on whether non-empirical or supraempirical factors can justify our denial that a competitor describes reality, add to "competitor" the notion of a *genuine* competitor: one *whose description of reality we cannot justifiably deny*. The competitor thesis from which issues of underdetermination arise is as follows:

> *For a given* T *there are genuine competitors*, distinct alternatives whose descriptions of reality we cannot justifiably deny.

The inference from such a thesis, against scientific realism, is that we cannot be justified in believing *T* merely because it factors into an abduction, that is, merely because it renders the phenomena a matter of course. With this it is important to emphasize that by no means is it the case that anti-realists who put forward competitors in the context of the scientific realism debate are putting them forward as candidates for belief. The antirealist is not in that business, or at least not in the business of believing claims one way or the other about unobservables.

Not only did J.S. Mill foresee the contemporary historical argument, as we saw in Section 1.3; he also anticipated a version of the competitor thesis (1867). Concerned with the nineteenth-century theory that light is a wave traveling through an ether, Mill noted that the mere fact "that an hypothesis of this kind" is such that "it accounts for all the known phenomena," does not, for "[m]ost thinkers of any degree of sobriety," justify our believing it. His concern was that "this is a condition sometimes fulfilled tolerably well by two conflicting hypotheses ... while there are probably a thousand more which are equally possible, but which, for want of anything analogous in our experience, our minds are unfitted to conceive" (1867, p. 296).[3]

One further question is that of what is required of the competitor thesis for it to threaten scientific realism. Traditional underdetermination arguments take it to be required that there are *empirically equivalent* competitors to *T*, where the competitors make all of the same predictions regarding observable phenomena. The threat in such a case is that irrespective of how much data we gather, we will never be able to distinguish between the competitors, so we will find ourselves

---

[3] As with various statements earlier, this is paraphrased to fit in the context of our current discussion, and it is only a preliminary glimpse. Nonetheless, it is worth foreshadowing that Mill's 1867 notion of hypotheses "our minds are unfitted to conceive" will be echoed by later philosophers (Duhem, 1906; Sklar, 1981; van Fraassen, 1989; Stanford 2006a, 2006b). No doubt, as with his historical induction, Mill's notion impressively anticipates the contemporary debate.

in a state of permanent underdetermination. However, the theories appealed to in the historical argument are *empirically inequivalent*, rendering as a matter of course phenomena *observed* at a specific time, but diverging with respect to at least some outside the observed subset; as testified by the historical argument, those empirically inequivalent theories nonetheless pose a serious threat to scientific realism. Turning back to the context of competitors, it is clear that Mill's point allows for either kind, and one question, which we will revisit, becomes whether empirically inequivalent competitors can also give rise to a state of permanent underdetermination.

We began in Section 1.1 discussing the realist's abductive inference that a given unobservable, U, exists. We introduced van Fraassen's challenge against the realist's demand for explanation and we had a glimpse at the historical argument. We then looked quickly at a third challenge to realism, the threat of a low degree of implication – an explanatory challenge relating to what Laudan calls the realist's "downward path." This challenge prompted us to explicitly recognize that what the realist needs in order to explain a given set of phenomena – and so what the realists, using their explanatory argument, claim we can justifiably believe – is not merely a posit that an unobservable U exists; it is instead, at least, a set of statements that attribute a set of properties to U, that is, a theory, T. With the recognition that the realist explanatory argument and commitment must extend to T, we are brought to a more sophisticated articulation of scientific realism than the one with which we began. And that sophistication opened the door for introducing a fourth challenge against realism: the competitor thesis and the threat posed by that thesis of the underdetermination of theories by data.

2 Believing the Best Explanation: The Realist's Move to Comparative Inference

2.1 Inference to the Best Explanation (IBE)

In Section 1.4, we recognized that abduction requires more than the mere statement that "U exists": minimally, having a greater degree of implication for observable phenomena–implying those phenomena to a greater degree – requires a theory that describes the unobservable entity. Similarly, but now with the competitor thesis in hand, we are invited to take an additional clarifying step that results in a further refined version of scientific realism, one even more sophisticated than the one on which we landed in the last section. This refined version of scientific realism can be seen as a preliminary defense against the competitor thesis we saw in Section 1.5; however, it will also invite two new challenges against realism. Specifically, we will see in Section 2.2 a fifth

Scientific Realism

challenge, one van Fraassen offers, namely, an alternative anti-realist description of the inference scientists employ. And we will see a sixth challenge to which we will give considerable attention, beginning in Section 2.3 and throughout to the end of Section 3: put briefly, at its core, the driving question in this challenge, and around which our inquiry will pivot, is, "What justification have we for believing that the non-empirical, or supraempirical, criteria that (realists claim) scientists favor in theory-choice have anything at all to do with reality?"

As a first step, toward a new sophistication of the realist position – and as a preliminary realist response to the competitor thesis in Section 1.5 – we are called to acknowledge that the realist is not committed to abduction as thus far construed. Rather, the realist adds an insight regarding the triadic nature of theory choice that became prevalent in the last half of the twentieth-century: scientific inference is not simply based on the relation between a theory and the phenomena; it is also based on a theory in relation to competing theories. Embracing this insight, our realist clarifies that it is not just any explanation – not just any theory that can be inserted into a Peircean abduction – that will do. It is the *best among competitors*. That is, the inference is properly construed, not merely as an abduction, but an Inference to the *Best* Explanation (IBE). While there are debates about whether abduction and IBE are equated, for present purposes I am treating them in such a way that IBE is the more sophisticated: it includes that twentieth-century insight of the triadic – and emphatically comparative – nature of theory choice. And with this we have a preliminary defense against the competitor thesis: when put up against the theories that scientists actually embrace, the competitors are inadequate; they constitute comparatively poor explanations.

In fact, on both sides of the debate, scientific realism is often construed as holding that we can justifiably believe the *best* explanation of natural phenomena. Clark Glymour expresses this sentiment: "One way to argue to a theory is to show that it provides a good explanation of a body of phenomena and, indeed, that it provides a better explanation than does any available alternative theory" (1984, p. 173). Peter Lipton writes that, from "a pool of potential explanations...we infer the best one" – "that the best of the available potential explanations is an actual explanation" (2004, p. 58), one that in fact describes underlying reality. Likewise antirealists such as P. Kyle Stanford express realism as embracing what I've dubbed the twentieth-century insight: "the justification" realists "offer for believing ... a given theory" is that "we think it offers the best available explanation for the empirical evidence we have and ... we regard rival or competing explanations of that same evidence as convincingly eliminated or discredited" (2006a, p. 122).

In addition to this "eliminative" aspect, as Stanford puts it, a crucial component of the justification of IBE in science – now carried forward from our discussion of abduction and van Fraassen's mouse – is that IBE is the mode of reasoning used in commonsense. Glymour writes, "sometimes the best explanation does go well beyond what is observed or observable ... the same features of inference which lead to general conclusions about the observable also lead in other contexts to determinate conclusions about the unobservable" (1984, p. 175). Lipton likewise writes that such "explanatory inferences are extremely common" (2004, p. 56) and sees IBE "as accounting for" not only scientific inference but also, importantly, "everyday inference" (p. 67). He writes, "many of our inferences, both in science and in ordinary life, appear to follow this explanationist pattern" (p. 1). He gives an everyday example: "Faced with tracks in the snow of a certain peculiar shape, I infer that a person on snowshoes has recently passed this way. There are other possibilities, but I make this inference because it provides the best explanation of what I see" (p. 1). So again we see the crucial appeal to our everyday reasoning. In fact, in van Fraassen's illustrative mouse example with which we began, it is IBE, in particular, that he is explicitly discussing.

On that note, we are led to a new critique from van Fraassen, one that challenges the empirical descriptive claim made by scientific realists that IBE, as the realist construes it, is the "rule" of reasoning in our everyday experience. It is this descriptive claim that is meant to license the thesis that IBE is the "rule" used in science; and that equation between the two domains of inference, in turn, provides a crucial element of the justification – arguably *the sole justification* – for believing our best scientific explanations that posit unobservables.

2.2 Challenge 5: A Descriptive Competitor to the Realist's IBE

Although realists will favorably cite the mouse example, we saw that it was introduced by an antirealist, van Fraassen. Moreover, after doing so, he contends that it serves the realist neither in a justificatory nor even merely descriptive capacity. Challenging the realist's descriptive thesis that *scientists use the realist version of IBE*, van Fraassen in turn challenges the realist's justificatory thesis, in short, the thesis that *we are justified in believing the best explanation* of a given set of phenomena, in particular those instances in which the best explanation posits *unobservables*.

As a preliminary note, and considering comments we saw earlier by McMullin and Devitt, although we have shifted to IBE, our realist is still describing first-order, or the base-level, scientific inference, which van Fraassen is likewise addressing. However, I suggest that van Fraassen is tacitly pointing out that the *description of inference* at the base-level is one step removed from that base-level inference itself. Van Fraassen calls the realist's claim a "*psychological hypothesis*" (1980, p. 20 [italics in original]). It is a level-2 *descriptive* hypothesis that we infer and so believe our best explanations, including, crucially for the realist, explanations invoking unobservables. Recognizing IBE as a level-2 hypothesis, van Fraassen himself invokes the triadic insight I specified previously: the realist description of commonsense and scientific inference, and the crucial component of equating the two, is itself "an empirical hypothesis" that is "to be confronted" not only "with data," but also "with rival hypotheses" (p. 20). And he challenges that realist's level-2 descriptive hypothesis, again about what is happening at the base-level, with a competing level-2 description.

Van Fraassen's "rival hypothesis" is this: "we are always willing to believe that the theory which best explains the evidence is empirically adequate (that all the observable phenomena are as the theory says they are)" (1980, p. 20). On this account, what we would believe is what the theory tells us about observables. Two clarifications are in order. First, van Fraassen is not prescribing or advocating his competing description; he is simply showing us that there is one. (For his own clarification on this, see his 1985, p. 295, ftnt 19.) Second, given the non-demanding nature of "adequate," "empirical adequacy" is too easily read – along with other phrases van Fraassen uses, such as "saving the phenomena" – as pertaining only to what will be, or even *has been, observed*. Common though this interpretation remains, it is incorrect. Van Fraassen writes, "I must emphasize that this refers to *all* the phenomena: these are not exhausted by those actually observed, nor even by those observed at some time, whether past, present or future" [original italics] (1980, p. 12). He likewise writes, "empirical adequacy goes far beyond what we can know at any given time. (All the results of measurement are not in; they will never all be in; and in any case, we won't measure everything that can be measured.)" (p. 69).

Adding to these clarifications, I propose that, ultimately, we can tie van Fraassen's rival hypothesis into the issue of underdetermination. That is, I suggest that van Fraassen is pointing out that the realist level-2 hypothesis – that we use IBE in commonsense and scientific inference – is itself underdetermined by the data. The claim that we believe the best explanation is ultimately empirically equivalent to the claim that we believe *something about* the best explanation, specifically, that it is empirically adequate. If van Fraassen's rival hypothesis succeeds, it knocks out the pivotal thesis that the realist's IBE is what we use in commonsense reasoning; doing so, it also knocks out the realist claim that we are *thereby* justified in believing conclusions about the unobservable.

Van Fraassen is showing that examples regarding mice and people wearing snowshoes fail to secure the first part of the descriptive realist claim, namely that we use the realists' version of IBE *rather than his rival* in everyday reasoning. The mouse and the person on snowshoes are, after all, observables. In those cases, the realist's claim that we infer their existence is identical in content to his rival hypothesis; with regard to observables, both hypotheses land on the same conclusion. It is true that, once we shift to the unobservable, they diverge in their content; however, the kicker here is that the two competing hypotheses nonetheless remain empirically equivalent. Van Fraassen is effectively claiming that, given his rival hypothesis regarding our inferences pertaining to observables, we are not compelled to go from IBE when the best explanation is something observable, such as a mouse, to the realists' IBE when that explanation invokes unobservables. It is empirically equivalent to say that in both cases we are believing that the best explanation is empirically adequate. Stepping out a bit, the realist is suggesting that, since we go from the pitter patter to the belief that a mouse exists whose feet pitter patter and whose teeth eliminate the cheese, we must, on pain of inconsistency, also go from a particular set of streaks in a cloud chamber photograph to believing the best explanation that a previously subnuclear proton affected by a magnetic field exists, and ultimately causes those streaks, and so on. The pain of inconsistency the realist attempts to demonstrate by appealing to our everyday reasoning is nonexistent: on van Fraassen's rival hypothesis, we believe that the theory regarding protons and their relation to the cloud chambers is empirically adequate, just as we believe that of the mouse explanation. Although the latter involves belief in the existence of something and its properties, and the former does not, no inconsistency is involved. We are pain-free.

Moreover, van Fraassen need not show that his rival is a *better* hypothesis. It need only be *as good, empirically*. It need only live up to what *it* demands.

That said, I add that it is not only natural but advantageous and warranted to re-invoke Challenge 1 in Section 1.2: van Fraassen's argument against the realist's demand for explanation. With that, not only do we lack grounds to prefer the realist's psychological hypothesis over van Fraassen's rival, we have reason to prefer van Fraassen's. In contrast with the realist's demand for explanation, accepting van Fraassen's hypothesis – that we believe that the best explanation is empirically adequate – allows us to block the infinite regress in a non-arbitrary and an epistemically relevant way. It has the virtue of posing no demand that leads to absurdity.

Nonetheless, here again it is important to emphasize that in putting forward a rival hypothesis, and even arguing for its superiority as I have, a non-realist need not, on, say, pain of inconsistency, concede to believing it. Even if van

Fraassen's rival is the best explanation for our inferential practices, lives up to what it demands, and allows us to non-arbitrarily block the realist's explanatory regress, I emphasize again my earlier point that the non-realist is not in the business of believing best explanations. (And even if van Fraassen were to partake in that business in terms of his own hypothesis, provided that his rival lives up to what it demands, i.e., is empirically adequate, that belief constitutes no inconsistency.) And finally, since realists use their description of our inferential practices at the observable level to *justify* their inferential practice regarding unobservable reality, in diffusing the realist description by introducing his rival hypothesis, van Fraassen has likewise pulled out from under realists (what appears to be *their entire*) justification for IBE. With a set of possibly disparate seeming points combined here, some cumulative strength is emerging against the realist's IBE and their claim that we can justifiably believe the best explanation. We will continue our watch to unveil other such tacit opportunities for dovetailing seemingly disparate challenges against realism.

2.3 Challenge 6: Supraempirical Criteria and Their Relation to Reality

Since the realist's justificatory claim depends on the realist's descriptive claim, van Fraassen's rival descriptive hypothesis strikes against both. However, we can also set aside that descriptive challenge and return to isolate and attend directly to the realist's justificatory claim, namely that believing the best explanation is *justified*.

To justifiably believe our "best" explanations about unobservables we are hard pressed to deny that we would also have to believe that underlying reality accords with those criteria humans just so happen to prefer in our attempts to describe it. But, the non-realist asks, what could possibly force the world itself to fit those "explanatory" criteria that we find we *like*, and so deem "best"? Or from the other direction, why should we take those supraempirical criteria toward which we happen to be inclined to be those that will pick out theories that actually describe, or to have anything at all to do with, reality itself – a reality that, as realists themselves emphatically hold, depends in no way on what humans happen to believe about it?

This challenge often arises in the context of the argument from underdetermination, introduced in Section 1.5. But as that argument has not been stated here, I suggest it is distinct. In fact, van Fraassen (1980, p. 90) on the antirealist side, and Lipton on the realist side, express versions of this problem in contexts where they are not explicitly discussing underdetermination. Looking to Lipton, he calls this concern, Voltaire's objection: "why should we believe that we inhabit the loveliest of all possible worlds?" (2004, p. 70). While I distinguish this challenge from the traditional underdetermination argument, it is nonetheless useful to explore it using competitors. In particular, I suggest that we look to competitors that *are not*, and we may *feel should not* be, taken seriously, those whose status as *genuine* is immediately denied. Such competitors strongly invite us to ask why they are not, and perhaps should not be, considered. And since that question brings out supraempirical criteria that *are* – and some that, it turns out, *are not* – at play in theory choice, it invites exploration into how we could justifiably believe that those criteria we happen to prefer allow us to *successfully* describe the physical world. Stated audaciously, who are we to demand of the universe that it does not instead accord with any other criteria, including the negations of those "explanatory" or supraempirical "virtues" we find ourselves preferring?

2.4 An Exploratory Tool: A Set of Exceptioned Competitors Extracted from Science Itself

Although plenty of lip-service is paid to the thesis that there are many potential alternatives to our favored scientific theories, that point tends to be hastily set aside. This is in large part because that topic of competitors is generally broached in the context of underdetermination, which, as noted earlier, tends to involve focus on empirical equivalence. In this section, I bracket both underdetermination and empirical equivalence. Instead, and to provide a tool for exploring the sixth challenge we've just seen, I will isolate and focus on what I dubbed in Section 1.5, "the competitor thesis." Specifically, I will endeavor to substantiate the claim that our favored scientific theories have many competitors.

I begin by offering a novel foundation for that thesis. Importantly, it is firmly secured within the content of contemporary science. According to contemporary science, each of the following statements (bulleted and in quotation marks) is among the very large class of empirically significant statements to which there are exceptions (some of which I indicate in the dashed portion beneath the bulleted statements):

- "Orbital speeds decrease in proportion to the square of the distance"
  — except in, for instance, galaxy NGC 3198, in which case the speed of many of its outer bodies does not decrease; that is, the galaxy's rotation curve is "flat."
- "Objects fall with uniform acceleration"
  — except when, for instance, an object is traveling through a gas at a high speed, has a wide cross-sectional area, and so on, in which case, it reaches a terminal velocity, that is, acceleration stops altogether, so is not uniform.

- "Light travels at 299,792,458 meters per second"
  — except when, for instance, it is traveling through media such as water, in which case it propagates at a slower speed.
- "Objects denser than water sink in water"
  — except when, for instance, the object is, say, an aluminum coin, which floats.
- "Processes in nature are reversible"
  — except in, for instance, entropic processes which are marked by their irreversibility
- "The pressure of a gas is inversely related to volume when temperature is constant"
  — except, when, for instance, carbon dioxide is under high pressure.
- "Chemical properties vary according to atomic weight"
  — except in, for instance, the case of some isotope-types where mass can vary but the chemical properties do not.
- "Subatomic particles obey the Pauli exclusion principle"
  — except when, for instance, the subatomic particle is a photon.
- "Strangeness is conserved"
  — except in, for instance, cases in which particle decays are mediated by the weak force and strangeness is not conserved.
- "Lepton-electron number is conserved"
  — except in, for instance, cases of neutrino oscillation. (I will discuss this example momentarily.)
- "Magnetite in basaltic lava emerging at divergent plate boundaries is magnetized in the direction of the North Celestial Pole"
  — except when, in some periods, for instance, the earth's magnetic field is / has been reversed.
- "Ice undergoing a heat induced phase transition turns to water"
  — except when, for instance, the molecules of glacial ice are under sufficient pressure and encounter sufficiently dry heat, in which case ice undergoes a phase transition directly into a gas/vapor, that is, sublimation obtains.
- "Genetic information flows unidirectionally from DNA to RNA to proteins"
  — except in, for instance, instances of reverse transcription in which case genetic information flows from RNA to DNA.
- "Agents of infection are living organisms"
  — except in, for instance, cases of transmissible spongiform encephalopathies in which case the agents of infection are prions; and
- "The chemical reactions in living organisms are catalyzed by proteins"
  — except in, for instance, a peptidyl transfer in which case the catalyst is a ribozyme, an RNA molecule, rather than a protein.

According to contemporary science, instantiating T with such statements (bulleted and within quotation marks) in the following expression and fully detailing their respective exceptions (of which, in most cases here I've only given singular examples, in the dashed portion beneath the bulleted statement), exceptioned theories of the following form will *better predict* and have better predicted the phenomena than each T in its non-exceptioned form.[4]

> *The world is as T describes, except in situation(s), S, in which case entity (or entities), E, will behave in manner, M.*

While I will discuss further implications ahead, we can note the following significant point here: taking the exceptioned theories I've listed to be predictively superior, as it does, contemporary science embraces *these exceptioned* theories *over* their non-exceptioned counterparts (bulleted and within the quotation marks). In fact, *the corpus of contemporary science entails indefinitely many such exceptioned yet altogether successful (and fully accepted) assertions.* Elaborating on an example in my list pertaining very clearly to unobservables, neutrinos are categorized as leptons (which have half-integer spin and, in contrast with quarks, are unaffected by the strong nuclear force). According to contemporary science, billions of neutrinos constantly barrage our bodies; yet the theory itself states, without question, that we have no way to ever observe that fact. Take "T" to be "lepton-electron number is conserved." By contemporary lights, *the world is as T describes except, in situation S,* for example, when a solar neutrino, which emerged from the sun as a lepton-electron neutrino, propagates through space, *in which case entity E,* the solar neutrino, *will* (often) *behave in manner M*; that is, it oscillates from its original state into another: it becomes a lepton-muon or lepton-tau neutrino. Given that, during its propagation through space, the neutrino has changed from one "lepton flavor," namely a lepton-electron neutrino, into another "lepton flavor," that is, a lepton-muon or lepton-tau neutrino, the particular situation, S, is one in which lepton-electron number *is not conserved.* So neutrino oscillation constitutes an *exception* to what we've just dubbed "T." While this example may sound obscure, these exceptions are of great significance in particle physics: neutrino oscillations are taken to have solved the multi-decade solar neutrino problem, where, in short, going back to 1964 neutrino detectors were taken to measure far fewer neutrinos, by a factor of two to three, than were predicted by the standard solar model. In fact, the 2015 Nobel Prize in Physics

---

[4] Bracketing irrelevant worries about the meaning of "laws," I take my usage of "exceptioned theories" to contrast them against Russell's reference to those "to which there are no exceptions." And for present purposes "exceptioned" and "non-exceptioned" appropriately capture the desired contrast. Directly contrasting more common terms leaves one or the other term insufficiently precise, for example, replacing "exceptioned" with "non-universal" or replacing "non-exceptioned" with "simple."

was awarded for the discovery of these oscillations; they are now accepted exceptions to the conservation of lepton-electron number.

Note that one is not restricted to relations asserted to obtain *in enormously many instances with just a few exceptions*; qualifying no less are the specification of relations asserted to obtain *in only a single instance and never otherwise*: in fact, one can just *begin with* the exception-clause ("*in situation(s), S, entity (or entities), E, will behave in manner, M*") and make *that* theory, *T*, and then add all the content of what had been *T* as the exception-clause. To sink our teeth into this point – again that one can specify *the exception-clause as T* and what *had been T* as the exception-clause – our lepton-electron example is telling. As noted, according to our contemporary corpus, lepton-electron conservation is *not* maintained in neutrino oscillation. Yet we can focus on the oscillation instead of the conservation, expressing the oscillation as theory, *T*: one can say "lepton-electron number is *never conserved*, except . . ." and then, in the exception-clause, detail all the indefinitely many instances in which lepton-electron number *is* conserved, according to contemporary science. One could do this, beginning with the exception-clause as *T*, even if there were just a single, or extra-ordinarily rare, exception to a given *favored T*: we can assert that extra-ordinarily rare instance as *T* and insert our favored theory as a dramatically extended exception-clause. Added to these two extremes are all the situations our contemporary corpus asserts to obtain in between. Given the utterly ubiquitous exceptions asserted by contemporary science, we would be gravely mistaken to expect that we can justifiably deny, and so reject, exceptioned theories – as, for instance, "implausible," "aberrant" (Maxwell, 1999), or "predictively unsuccessful" – *solely because they take this exceptioned form*. Emphatically, again, the exceptioned theories in my list are extracted from, and taken to be the case by, contemporary science itself.

We now have a novel foundation for *the competitor thesis* (Section 1.5) drawn from our own scientific corpus: the recognition of *predictively successful exceptioned* statements that *are accepted* by contemporary science. With this foundation, we can now step outside of the contemporary corpus and ground the thesis that our favored scientific theories within it have many competitors. To see this, take now any *non-exceptioned theory* we may favor, whose tested predictions have been observed thus far to obtain, *without exception*. That non-exceptioned theory could be, for instance, a deeper-level theory taken to *account for* one of the accepted exceptioned theories pointed to in my list. Inserting now for *T* in the indented expression (just after my list) any *favored non-exceptioned theory* (and remaining mindful that the expression is one that is solidly exemplified in the corpus of contemporary science), it is clear that there are exceptioned *competitors* that predict the same *observed* phenomena predicted by that favored non-exceptioned T. For this set of competitors, S can be any set of circumstances that has not yet been confirmed as occurring, the possibilities for S being inexhaustible: an interaction between specific entities or forces, a condition or process that could be brought about in a yet-to-be-attempted experiment, a yet unexperienced spatiotemporal location, and so on. And any entity or force, observable or unobservable, whose behavior T describes can qualify as E. M can be any conceivable behavior that is significantly distinct from the behavior predicted by our favored T, so that T will be unequivocally contradicted by M. We recognize that indefinitely many options are available for each of the variables S, E, and M. And we can hardly exhaust the ways in which the particular S's, E's, and M's can be mixed in the exception-clause.[5] The point here, of course, is that there are competitors that assert an exception-clause that S, E, and M obtain. Granted – unlike the exceptioned theories drawn from within accepted science, for instance, examples in my list – the empirical data do not favor these new exceptioned *competitors* over our favored non-exceptioned T. *But crucially, nor do those data favor what we happen to favor, our non-exceptioned T, over its exceptioned competitors.*

Moreover, and again, but this time with respect to *competitors to our favored T*, one can just *begin with* an exception-clause (as in our example earlier, neutrino oscillation), making it the theory, T, and turn theory T (e.g., as earlier, all instances of lepton-electron conservation) into the exception-clause. We need only extend this same exceptioned form to any other claim about unobservables that happens to be, by present lights, taken to be non-exceptioned: make our favored non-exceptioned theory, T, itself constitute an exception-clause, and designate as T a statement of exceptions to that favored theory.

Acknowledging that there are competitors that assert that S, E, and M obtain is of course entirely distinct from asserting anything along the lines that a particular articulation of S, E, and M in fact obtains. The non-realist is patently not an epistemic realist about competitors. In short, any given non-exceptioned T we may favor is *empirically undistinguished* from indefinitely many exceptioned competitors. In fact, since there will always be indefinitely many, even if a few can be empirically eliminated, our favored T is always empirically *indistinguishable* from indefinitely many competitors.

As noted in the last section, and now with a set of competitors revealed by the expression that follows my list, we can use that expression as a tool to explore the criteria on which those competitors may, descriptively at least, be commonly

---

[5] Three points: First, for some theories, every object is included in the class of entities, E, to which it pertains. Second, none of this should be taken to imply that, for E, only entities described by T qualify. Third, although E here denotes entities and M their behavior, E can also denote processes and M their occurrence.

ignored. This will bring us into direct contact with several supraempirical "virtues," such as simplicity, explanatory breadth, and even coherence with background theoretical systems, and will allow us to explore what grounds – in the realist attempt to describe reality – could possibly justify our disregard of theories that fail to live up to those virtues. Mindful of that, we can frame our inquiry with the following question:

> How can we *justify denying* that the indefinitely many exceptioned competitors describe reality and claim *justified belief instead* for our favored non-exceptioned, and *empirically undistinguished*, theories against which the former compete?

2.5 Some Pre-Reflective Exclusionary Proposals

Let us first address pre-reflective reactions on which we might think we can base the dismissal of those competitors as descriptions of reality. Of course, because each will share T's empirical success so far, we simply cannot discard them for failing to fit the data. And since, irrespective of future empirical tests, there will always be indefinitely many such competitors; the *set of competitors* can never be eliminated *empirically*. (Our *primary* motivation for revealing this exploratory tool is Challenge 6, introduced in Section 2.3, "supraempirical criteria and their relation to reality." We are not, currently, explicitly concerned with addressing "the underdetermination of theories by data," per se, Challenge 4, introduced in Section 1.5. However, it is noteworthy that, since the set of competitors can never be empirically eliminated,[6] we have landed on a state of *permanent* underdetermination – despite the fact that we have not been concerned with empirical equivalence, which our inquiry now suggests is a gratuitous demand.) The fact that we can never empirically eliminate the set of competitors reinforces the notion that any exclusionary criteria must be *supra*empirical, potentially identifiable in advance of any empirical test; and it at least suggests that those criteria will pertain specifically to the theory's *form* – where, for instance, T's form is such as to include an exception-clause; that is, T takes an exceptioned form (as captured in the expression in Section 2.4 derived from science and repurposed to reveal competitors); or T has the form of including no such clause; that is, T is non-exceptioned.

---

[6] Bracketing my use of the corpus of science as a foundation for the competitor thesis, I am indebted to Peter Lipton (1994), Richard Swinburne (1997), and Nicholas Maxwell (1999) for helping me see the problem of what I am calling "exceptioned competitors." Nancy Cartwright has a robust research program (1983, 1999) that contains as a core element the idea that the fundamental laws of physics insofar as they hold at all, do so only in very rare, and sometimes entirely idealized, circumstances. My discussion of idealizations in my (2017), and exceptioned theories here, especially at the deepest level, has a clear relation and debt to her research program.

Of course, on that note comes a second pre-reflective response: with no distinguishing data, we may find in ourselves a strong psychological distaste for even thinking about the exceptioned *competitors*. Descriptively, at least, we do not take them seriously, generally discarding them as uninteresting or even absurd in advance of even considering, let alone testing, them. When pressed on why, we might say they clash with "commonsense" or might deem them counterintuitive. However, first, "commonsense" and the "intuitions" of given periods have been defied by many exemplary theories, our very best, relativity and quantum mechanics, among them. And since, as emphasized in Section 2.4, science embraces many exceptioned theories *over* their non-exceptioned counterparts, merely deeming the form of the competitors counterintuitive or even psychologically unbelievable will not legitimate denying that they, rather than our preferred theories, describe reality. Appealing to such "tastes" we may happen to have will clearly fail to suffice in the context of present concern.

Another more tenable exclusionary proposal might be that the competitors are parasitic on theories scientists actually embrace, and that theory parasitism justifies denying that exceptioned competitors describe reality. Fleshing out this notion of theory parasitism, one might protest that the competitors embrace some, but deny other, aspects of another scientifically legitimate theory. However, first, this is exactly the case for numerous theories accepted in the history of science; one might dare say that most have had that kind of theory parasitism, precisely embracing some portion of another theory while nonetheless denying other portions: Kepler's did this to Copernicus's; Newton's to Kepler's; Clausius's to Carnot's; Hesse's did this to Wegner's; Einstein's photon theory did this to Maxwell's, as did his special relativity; Gamow's did this with Lemaitré's cold primeval atom; and Guth's inflationary theory did this to Gamow's hot big bang, and so on. Elaborating a bit on one example here, Newton very clearly used Kepler's three laws of planetary motion in his (1684), (1687). However, he nonetheless wholly discarded Kepler's theory of the *anima motrix* to which Kepler was deeply committed and which, most importantly, Kepler genuinely used to arrive at those laws Newton used. And beyond that, as Duhem points out (1906), the Newtonian system once worked-out predicted non-Keplerian perturbations. Newton embraced a selected portion of Kepler's theory on which Newton built his own theory, while wholly rejecting the core components of Kepler's. (For more detail see my 2006b.) Although I've included the charge of theory parasitism as a pre-reflective reaction, it is, surprisingly, one that is seriously entertained in the realist literature – despite the fact that the history of science is replete with such instances of "theory parasitism," namely, fully embracing some, but wholly discarding other, aspects of another, for example, previously held, theory.

Scientific Realism 23

It is worth pausing briefly to note that the exclusionary criteria we have just now set aside can be rejected on descriptive grounds, which we will also see in nearly every example to come. When it turns out that exclusionary criteria would also preclude theories that have been clearly embraced within science, the scientific realist, on pain of denying that scientific theory choices are justified, simply cannot claim justification for those exclusionary criteria. So not only does our discussion hold promise for bringing clarity to criteria that are at play, it also holds promise – as I briefly hinted in the last paragraph of Section 2.3 – for bringing to the fore criteria thought to be at play in scientific theory choice but, upon examination, we discover, are not.

Continuing with pre-reflective reactions, one might simply point to the fact that no scientist entertained or at least explicitly proposed such competitors. In response we might cite historical challenges to this claim – for instance, that with the discovery of beta-decay, Niels Bohr proposed that beta-decay stands as an exception to both the conservation of energy and of momentum. However, we can more fundamentally return to our starting point – second paragraph, Section 2.4, containing our list – that the corpus of science is filled with such exceptioned theories. We can add that one would be hard pressed to show that none of those exceptioned theories were proposed on their own before some deeper non-exceptioned theory could account for the exceptions. Beyond such descriptive concerns, even if it were the case that scientists have never proposed exceptioned competitors, invoking that to ground the latter's exclusion is to misconstrue the role of the expression of the competitors and our current purpose: the claim is not that reality corresponds to any particular competitor scientists might actually articulate against a given favored theory; the role of the expression is rather that of *revealing* that there are indefinitely many exceptioned competitors. And again, should one complain that no data distinguish some exceptioned theory over our favored non-exceptioned theory, we reiterate that, because our favored theory is empirically undistinguished from each of the competitors, that complaint has identical strength in each direction: no data distinguish our favored non-exceptioned theory over its set of indefinitely many exceptioned competitors. Clearly, something more is needed to exclude the latter.

On that note, one might protest against the method of attaining these competitors. However, first we return to the crucial fact that their attainment was *based on* statements contained within our contemporary corpus. Second, we reiterate the fact just noted, that this misconstrues the point of the expression: it is to *reveal* that there are always indefinitely many exceptioned competitors rather than to propose or generate any specific competitor. Third, even if the expression were employed to *generate* a set of the indefinitely many competitors, we could reject those generated neither for that alone, nor for failing to be "derived" from our favored theories or "directly induced" from observations. It is now wholly granted – with a hearty wave to the likes of Popper and Hempel – that strict rules for theory attainment have not been upheld in the history of science: any number of accepted theories have been generated by any number of means, including guesswork, dreams, religious grounds, and hallucinogenic drugs. Mendeleyev, Kekulé, Bohr, for instance, claimed to have dreamt up the periodic law, a model of the benzene molecule, and a model of the atom, respectively. Kepler's laws were strongly based on his mystical Neoplatonic-Pythagorean-Christian conception of reality. And at least one Nobel Prize winner claimed to use LSD in his key discovery, namely, the "invention of the polymerase chain reaction (PCR) method" (Mullis, 1993).[7] He is said to have asked, "Would I have invented PCR if I hadn't taken LSD? I seriously doubt it ... I could sit on a DNA molecule and watch the polymers go by. I learnt that partly on psychedelic drugs" (Nichols, 2016, p. 332).

We've now seen that scientific realists modify the inference to which they appeal, moving from abduction, which makes no reference to other theories, to its comparative cousin, inference to the best explanation. This necessary realist sophistication invites two new challenges against it, beyond the four introduced in Sections 1.2–1.5. Specifically, we saw in Section 2.2, a fifth challenge: a rival anti-realist description of the inference that scientists employ, namely, that they infer only that the best explanation is empirically adequate. We then arrived at a sixth challenge in Section 2.3, essentially captured in the question, "What justification have we for believing that the supraempirical criteria (realists claim) scientists favor in theory-choice have anything at all to do with reality?" In need of a tool for exploring this question, we took a novel approach to revealing competitors. We first extracted the notion of exceptioned theories from the content of contemporary science itself. We then stepped outside of accepted science, and using the form of these exceptioned theories, we repurposed it and introduced an *empirically ineliminable set of exceptioned competitors*. We've also now seen that realists cannot justifiably reject those competitors by invoking the pre-reflective exclusionary criteria we've considered here, for example, on the grounds that they fail to fit the data, that they conflict with commonsense, that they are counterintuitive, that they are parasitic on the favored theories to which they are competitors, or that they defy some rule for theory attainment. We will now turn to consider a set of ostensibly

---

[7] I thank my former graduate student, Monica Morrison, for alerting me to this case. The general point I'm flagging here is often discussed in terms "discovery versus justification," but I find the phrase "theory attainment" to be far more direct than "discovery," which can pertain to so many scientific achievements beyond theorizing, for example, the discovery of an unexpected phenomenon. I also find it misleading to say theories are "discovered," especially when they can later be taken to be false.

Scientific Realism 25

more tenable exclusionary criteria, supraempirical virtues, to which realists may or do, in fact, appeal in want of excluding competitors.

3 Supraempirical Virtues and Their Prospects for Justifiably Excluding Competitors

3.1 Virtue 1, an Overarching Virtue: Supraempirical Criteria Are Justified Provided They Are Inherited from Empirically Successful Background Systems

According to Stathis Psillos (1999) – nodding to Richard Boyd (e.g., 1973) and Wesley Salmon (e.g., 1985) – a theory gains its evidential support not only from its empirical success, but also by its relation to background theories. There are two proposals for excluding competitors based on a theory's relation to background theories. One is that *coherence with the background system in place* is itself a virtue capable of excluding competitors. I'll address that proposal in Section 3.2. The other proposal, which I'll now address, is an attempt to justify the appeal to *supraempirical virtues in general* by the fact that *those virtues* – whatever they may turn out to be – *are present* in background theories. As to why, from a realist perspective, the fact that background theories exhibit a given virtue provides justification for favoring theories that also exhibit that particular virtue, Psillos explains:

> These background theories have themselves been accepted because they enjoyed evidential support and displayed similar virtues. Hence, their evidential support and theoretical plausibility are carried over, and reflected in, the new theories which they license. The virtues which constitute explanatory power become evidential [i.e., provide evidential support] precisely because they are present in theories which enjoy theoretical plausibility and evidential support. (1999, p. 172)

First, this proposal, even if successfully implemented, does not appear sufficiently potent to justify the *exclusion* of competitors. Second, it only pushes the question further: why should these background theories themselves be attributed "theoretical plausibility" in their capacity to describe reality. The initial answer: not only because their empirical predictions are confirmed, but also because of their explanatory virtues. The question of why the virtues should be considered evidential in the first place remains. Why should we think they make the background theories (and thereby our favored non-exceptioned theory) more likely to properly describe reality? Psillos answers: "Given two theories T and T' which have the same observable consequences[8] but are differentiated in

---
[8] Psillos is talking about empirical equivalence, with which our topic is not concerned.

respect of some theoretical virtues, one should regard T more plausible than T' if, given the past record, theories which exhibit the virtues of T are more likely to be true than are theories like T'" (p. 172).

Three points stand out. First, Psillos is proposing that we test the virtues *against the historical record* of scientific theories. Second, rather than offering results, he is *merely proposing a test*. Third, *Psillos's proposal, as just stated, is clearly not going to work*. His broad goal is to show that we have a general way to determine whether a theory is "likely to be true." His particular approach is to attempt to establish that the presence of certain virtues will make a theory "more likely to be true." The method he is proposing to establish this: identify a correlation between (a) theories that have particular virtues and (b) theories that are "likely to be true." Assume the particular virtues are such that we can identify instances of a theory possessing them, so instances of (a). Nonetheless, without already possessing a way to show that a theory likely describes underlying reality – that is, without proclaiming a realist victory in advance of the realism debate – we cannot identify a single instance of (b). We have traversed a futile circle. Psillos's method for solving the problem, for addressing an issue in the realism debate, requires that the problem has already been solved, in fact that the entire realism debate has been resolved with victory going to realism. I dare say there can be no content to this approach as phrased.

Giving Psillos the benefit of the doubt, however, let us excuse his use of "true" and "likely to be true" as slips. In the next sentence, intending to provide an example of the approach just considered, he appeals to something weaker, namely evidential support: "So, for instance, if theories which have not been subjected to *ad hoc* adjustments have tended to be better supported by the evidence than theories with *ad hoc* features, then this consideration should be used in assessing the prior probability of other theories, in order to rank higher theories with no *ad hoc* features" (p. 172). Thankfully the correlation between certain virtues and *evidential support* can, in contrast with his initial proposal, be historically/empirically assessed. Testing this correlation empirically, we would discern whether theories with a certain virtue have "tended to be better supported" evidentially.

Unfortunately, however, for Psillos, we are caught – or at least must take careful measures to avoid getting caught – in a second though less blatant loop. On his way toward his broad goal of showing certain virtues indicate that we are genuinely describing reality, Psillos is trying to show that there are certain virtues which, in addition to empirical confirmation of predictions, provide evidential support: "The virtues which constitute explanatory power become evidential" (p. 172). The method meant to establish this is to show that theories with certain virtues are well-supported. That given, in testing this proposal,

Scientific Realism 27

"evidential support" cannot include the possession of any virtue we are seeking to correlate with evidential support. For that would amount to saying that T not only has the virtue of, say, being non-exceptional, it also has evidential support in the form of having that very virtue. The correlation would be tautologically true thus vacuous. So – *in these particular tests at least* – empirical support is the sole type of support that can count as evidential.

Having clarified this, we will ask whether theories possessing a set of virtues have "enjoyed" *empirical* "support" (p. 172). While Psillos is promoting this for a potential range of supraempirical virtues, others such as Alan Musgrave (1985, pp. 203–4) have also espoused this approach for *particular* virtues. Neither has shown us a positive result. Both have only assumed the history of science would reveal the desired correlation between a certain virtue and empirical support. Even supposing it did, such a two-placed correlation will not be sufficient. We will also need to see whether there are theories with those virtues that do not enjoy empirical support and whether theories that lack them do.

Here I offer a promissory note that I will return to this issue of empirical support for theoretical virtues, specifically simplicity, in Section 3.6, where we will find ourselves compelled to draw radically more negative conclusions. For now let the following suffice: even if we implemented the (merely) proposed test and successfully distinguished "the virtuous" among theories as faring better empirically than the wretched, that would do nothing to show what Psillos and Musgrave hope, and, as scientific realists, minimally *need*, to show, that our attempted descriptions (of underlying reality) possessing those virtues relate *in any way at all* to underlying reality itself. Our (assumed to be) successful test would only show that theories possessing those virtues "enjoy" empirical "support," a correlation happily welcomed by an empiricist. Connecting the virtues only to empirical support does nothing to tie them to underlying reality, and, it appears, does nothing for scientific realism per se.

While this empirical method for justifying virtues *in general* does not look promising, there may nonetheless be specific supraempirical virtues whose justification is distinct. As noted at the end of Section 2.4, our driving question is,

> How can we *justify denying* that the indefinitely many exceptioned competitors describe reality and claim *justified belief instead* for our favored non-exceptioned, and *empirically undistinguished*, theories?

Seeking, on behalf of the realist, to eliminate the threat posed by exceptioned competitors, we are searching for some supraempirical restriction(s) that

1) we can justifiably assert to bear on our ability to describe reality
2) will genuinely preclude exceptioned competitors from being included in the class of *genuine* – reality relevant – competitors.

I will now explore the prominent contenders.

3.2 Virtue 2: Coherence with the Background System in Place

Though a set of problems emerged for that general proposal pertaining to background systems, with which we concluded the last section, there is a second distinct, more common, and more direct way realists invoke background systems. Competitors are excluded not by virtues T *inherits* from background systems, but by background systems themselves. T's relation to, its coherence with, the background system is itself the virtue that competitors lack. Realists commonly invoke this explicitly as a requirement to eliminate competitors (e.g., Lipton, 1993/2004, pp. 157–63; Psillos, 1999, pp. 217–19). Because the background system has in itself attained its own empirical support, realists take it to meet condition (1), we can justifiably claim it bears on our ability to describe reality; and anticipating that competitors will fail to cohere with our background system, realists expect this demand to also meet condition (2), to exclude exceptioned competitors. That is, requiring a theory to cohere with the background system appears prima facie to both bear on reality and justify discarding exceptioned competitors. Without question we have before us more challenging proposals than the pre-reflective reactions in Section 2.5.

However, we began this inquiry in Section 2.4 by extracting exceptioned competitors from contemporary science – which, I emphasized, is ubiquitously *filled* with exceptioned theories. Recognizing further that every non-exceptioned theory in that corpus can also be replaced with an exceptioned theory, it is clear that there are innumerably many broad-ranging, yet profoundly exceptioned, systems with which a given exceptioned theory or set thereof can cohere. Another way to put this is that our favored background systems are empirically undistinguished from and so threatened by exceptioned competitor-systems, no less than is a favored non-exceptioned theory.

Moreover, this makes especially clear that condition (2) is met only with the further restriction that *the background system is the one already in place* (hereafter, "background-in-place").[9] However, endeavoring to describe underlying reality, we have no grounds for that demand. In terms of background-coherence, condition (1) leads us only to a less demanding requirement: our

---

[9] Realists liberally use "background knowledge." However, to avoid misleading ourselves – even by tacitly, granting victory to realism in advance of the debate – I have replaced that term here with this appropriate phrasing.

theory must cohere with a set of statements that collectively account for a broad range of data. The background-in-place may be one set that suffices, but it is patently not the only set. The less demanding restriction provides no more and no less than what condition (1) requires. Anything more – such as the background-*in-place* demand – is superfluous.

Importantly there is also a serious descriptive challenge to the insistence that our theory coheres with the background-in-place: it is defied by the history of science. Were that demand – to which realists tend to pay such heavy lip-service – held to, it would have prohibited, for instance, small scale modifications of the kind we see in the three decades of steps toward quantum mechanics, for example, Planck's, Einstein's, Rutherford's, Bohr's, and De Broglie's steps. Each blatantly contradicted and defied classical background posits. To take one example: in light of the latter posits, including Maxwell's extremely successful theory of electromagnetism, Rutherford's posit of an atomic nucleus around which electrons orbit predicted that electrons would lose their energy and within a millisecond crash into the nucleus, resulting in the elimination of all objects. Despite this radical conflict between Rutherford's posit and Maxwell's theory, Rutherford's posit was not rejected. Moving forward historically and more generally, quantum mechanics did not cohere with Newtonian determinism. And outside of quantum mechanics, special relativity failed to cohere with the theory of the luminiferous ether. This is so despite the fact that no one in history had identified a single context in which waves could travel without a medium – be it ocean waves, sound waves, and so on. James Clerk Maxwell himself insisted on such grounds that "there can be no doubt" (1890, p. 775) "that there must be a medium" through which light waves are propagated (1873, p. 438). Laudan paraphrases Maxwell's comments: "the aether was better confirmed than any other theoretical entity in natural philosophy!" (1981, p. 27). Despite the fact that many, Kant being only one among them, took Euclidean geometry to be an indubitable description of space, general relativity failed to cohere with it, nor even with the claim that there is an instantaneous action-at-a-distance gravitational force. Paul Thagard (1992), though a realist advocate of IBE, is compelled to discard the background-in-place demand. Regarding the examples just noted, he points out that the following were all defied by relativity, with "the first three ... eliminated by the special theory alone" (1992, p. 209):

(1) Time and space are absolute;
(2) There is a luminiferous aether;
(3) Objects have no maximum velocity;
(4) Euclidean geometry adequately describes space;
(5) There are instantaneous gravitational effects;
(6) Light travels through space in straight lines. (p. 209)

Likewise, by no means could Newtonian mechanics be merged with the underlying reality as described by the Aristotelian background system. As Thagard notes, "Copernicus, Galileo, Newton together overturned the entire Aristotelian system." Specifically, "Copernicus rejected the spatial arrangement, moving the earth from the center of the universe, and he altered the kind relations concerning celestial bodies, installing earth as a kind of planet" (p. 192). Galileo's "principle of inertia abolished the distinction between natural and unnatural motion and rest," while "Newton finally eliminated the distinction between celestial and terrestrial bodies, showing them all to be subject to the same laws of motion" (p. 193). Of course, these comments only hint at the radical overthrow of the background-in-place brought about by these revolutions.

No doubt each of these is an exemplary case of theory change that the scientific realist must concede to be progressive. In fact, demanding background-in-place coherence to get at underlying reality, realists would be left unable to account for these exemplary transitions. That casually but so commonly expressed demand would prohibit each one. Restricting our focus for now on background coherence (bracketing other supraempirical criteria to be discussed in Sections 3.3–3.6), the requirement these exemplary theories had to meet was the one we landed on previously: eventually, with a background system, they accounted for a wide range of data. Empirical success aside, most new broad-ranging theories bring their own, entirely new, background system. Like Thagard, Richard Swinburne is a realist who rejects this background-in-place constraint: "when we are considering very large-scale theories," he writes, "there will be no such background knowledge" (1997, p. 37). After noting this for Newton's theory in his (2001), he adds "and the same point applies, even more strongly, to the even more all-embracing theories developed since Newton that seek to explain all things known to science" (2001, p. 93). Where elaborating, he writes, "we have now theories of the weak nuclear force and of the strong nuclear force, as well as Quantum Theory. And a Theory of Everything, like superstring theory, seeks to explain all things now known to science" (1997, p. 37). Because such theories are so far reaching, there will be no "theories of neighbouring areas with which such a theory could dovetail" (1997, p, 37; see also 2001, p. 93). "A theory of everything," for instance, "does not have to answer to any background knowledge" (1997, p. 40). In contrast with other virtues, coherence with the background-in-place fails to qualify as a property that is or *even should* be demanded of theory selection.

The general *prescriptive* point on which we are landing is classically emphasized by Paul Feyerabend (1963).[10] (See also Khalifa (2010).) Recently, K. Brad

---

[10] Of course, favorably embracing this particular empirical point does nothing to commit one to embracing Feyerabend's more radical claims.

Wray (2018) adds examples where the theories are less fundamental to reality overall, we might say, but nonetheless fundamental to a specific domain. He writes, "geologists working with the assumption that the continents are fixed are unlikely to entertain or develop hypotheses that ascribe motion to them," and "if the continents do, in fact move," as our contemporary corpus tells us, "such background assumptions will be an impediment" (2018, p. 62). He adds another example, contrasting Newtonian theory against, not Aristotelian but contemporaneous theories. Those who worked with "the assumption that all motion is due to contact between bodies," for instance, Cartesians, were "unlikely to develop a theory according to which there is action at a distance" (p. 62). And if in fact the processes in the world act by way of such action, those scientists will impede "the advancement of science" (p. 62). As Wray puts it, if background systems "do narrow scientist's thinking, the narrowing is not necessarily going to have a positive effect" (62–3).

Facing a history of background systems being overthrown, our realist cannot justifiably impose onto reality a background-in-place demand. It fails to capture a property that is, ought, or can be required in theory selection. It is not merely superfluous, but historically defied. Forced to set that demand to the side, the realist endeavor to get at reality can require no more than coherence with a set of statements that accounts for a broad range of data. We cannot then preclude exceptioned competitors merely for needing their own background system, even assuming they do. Returning to our earlier point, indefinitely many sets of exceptioned auxiliaries can be constructed to constitute new background systems, resulting in systems against which our own is empirically undistinguished. Surprisingly, a coherence-with-background mandate appears incapable of – fails as a candidate for – excluding the exceptioned competitors.

3.3 Virtue 3: Indirect, Vicarious Support

As an anti-realist, Laudan does not claim we can justifiably believe our best explanations, but only that theory choice is rational. Nonetheless, in collaboration with Jarret Leplin (1991, 1996), a realist, he argues that we can often distinguish between competitors by way of support that does not arise from a theory's own consequences. A statement can inherit the support attributed to another when both are entailed by a more encompassing statement. While "the next crow to be sighted will be black," does not entail "past sighted crows were black," the former receives *support* from the latter. This arises from the fact that each is entailed by a broader universal claim. Statements, they argue, can receive vicarious evidential support. This extends to broader ranging theories as well. Say of two competitors, H1 and H2, that "H1, but not H2, is derivable from a more general theory T, which also entails another hypothesis H" (1996, p. 67). Since H is entailed in T, a consequence of H will support T. Doing so, it will indirectly support H1. (See also Lipton, 1993, p. 65; 2004 p. 63.) Let us grant that a theory can receive such indirect support.

Even so, earlier considerations apply here. First, since there will always be some broad theory that is not subsumed by another, one obviously cannot *require* that a theory receive this form of support. In fact, just as most large-scale theories will not conform to any *background-in-place*, theories like Newton's or Einstein's will not be entailed by any available theory so can receive no such support. Laudan and Leplin's points will be altogether inapplicable at this scale. Second, just as the retention of the background-in-place cannot be required of a description of reality, the latter does nothing to require retaining the entailing theory exactly as it is. That given, we realize that there are subsuming competitors that will entail a given exceptioned theory and from which the original subsuming theory will be empirically undistinguished. The original subsuming theory need only be made exceptioned by one among indefinitely many potential exception-clauses. Because we cannot require of a given theory that it be entailed by another, and because, even if we did, it would not prohibit exceptioned theories, the possibility of indirect, vicarious evidence as depicted in this section fails to block our exceptioned competitors.[11]

3.4 Virtue 4: Breadth of Scope

There is little doubt that breadth of scope is among the desiderata of scientific theories. The idea is that a theory with greater scope is favored over one that can account for the same data but has less scope. Kepler's theory accounted for planetary motion to an unprecedented degree. Galileo purported to account for the motion of many objects on Earth. Newton's theory, encompassing both domains, purported to describe all objects in the universe. Considering breadth of scope as a criterion desired in our explanations, we can properly dub it supraempirical in that it goes beyond the data that have been observed at a given time. This virtue contrasts with the three thus far considered: because breadth of scope is agreed to be strongly favored in science, it is far more difficult to deny that it is required of theories. Let us accept that it is.

Even then our primary questions remain: is the demand for breadth of scope a criterion the exceptioned competitors fail to possess, and, assuming they do not possess it, can we justifiably deny that they describe reality? Beginning with

---

[11] Leplin and Laudan also contend that "sophisticated analogies can be evidentially probative" (1996, p. 67). However, in agreement with an anonymous referee, I've eliminated discussion of this virtue due to the obviousness of its failure to exclude competitors as descriptions of underlying reality.

the latter, it appears we cannot. On the contrary, as is well known, a broader statement claims more than a narrow statement it entails; it has greater content that exceeds its empirical support. Therefore, the broader statement can fail to describe the world in more ways than the narrower statement. Because the latter says, entails, implies less, it has greater logical probability than the broader statement. In van Fraassen's words, "credibility varies inversely with informativeness" (1985, p. 280). A claim that fits yet goes beyond a set of data will be less likely to obtain in the world than one that is empirically undistinguished from it but has less breadth. Hence any narrow, exceptioned theory that fits the data has greater logical probability than any competitor we favor for its breadth. Given the indefinitely many options for exceptioned statements and systems, many exceptioned systems as a whole will have just as great, if not a greater, logical probability than our preferred system. While breadth of scope beyond what's been observed is a key explanatory property desired in theory choice, it cannot justify the denial that an exceptioned competitor properly describes reality. (See also Laudan, 2004.)

Further, let us momentarily assume, contrary to logical fact, that greater breadth of scope increases the chance that we have properly described the world. Although we have just noted that there are exceptioned theories with less breadth, because again there are indefinitely many exceptioned theories, nothing in the realist armory bars exceptioned theories whose scope is equal to that of their non-exceptioned counterparts. The exception-clause can imply just as much about what has not been observed as can the same portion of the non-exceptioned theory. So even neglecting the fact that breadth of scope makes a theory less likely to describe the world, we find ourselves without a way to preclude the exceptioned competitors. We cannot block them for failing to have the same breadth of scope as our empirically undistinguished non-exceptioned theories; nor have we any justification for denying that those with less scope genuinely describe reality.

3.5 Virtue 5: Novelty

We are seeking some criterion realists can justifiably invoke that blocks exceptioned competitors to our favored, yet empirically undistinguished, non-exceptioned theories. We will later see that novel predictions become an integral part of the realism debate. For now, we note that the realist might attempt to invoke them to exclude the exceptioned competitors.

However, unless we are willing to suggest that scientists theorize with no reason for doing so, we can hardly claim of *any* historical theory that its author was not mindful of, and so using, some data in conceiving and developing the theory. It is implausible that any theory – including any non-exceptioned theory we may favor – that failed to accommodate any data would have been taken seriously by its author, let alone the scientific community. And, because no one can deny that accommodating theories can describe reality, even if an exceptioned competitor made no novel predictions, that would provide no grounds to deny that it describes reality. Beyond that, though, even a superficial glance reveals that we can direct no such general charge against exceptioned theories: the exception-clauses themselves will often imply, possibly directly, a range of distinct novel predictions. For this reason, likewise, falsifiability or "independent testability" mark no excluding factors.

The strong demand of course would be for *confirmed* novel predictions. As this pertains to the empirical realm, it is worth noting we have now shifted to consider a virtue that would *not* be *supra*empirical per se. Even so, and even if confirmed novel predictions were required of theories in science, such a demand would not block the exceptioned competitors. Because the expression we used to reveal exceptioned competitors reveals them for all theories, it does so for those attaining novel success. Moreover, since the realist insists, emphatically, that reality is what it is irrespective of what humans might think about, write, believe, and so on, the mere fact that a competitor may be unarticulated remains wholly irrelevant. Importantly, the exceptioned theories stand as competitors at the very moment a favored theory is devised and hence at any subsequent moment a given novel success is attributable. Furthermore, irrespective of whether exceptioned theories are ever articulated, their status as competitors obtains prior to any confirmation of the novel predictions of the non-exceptioned theories we may favor. Among the full class of competitors that do not diverge in respect to a given successful prediction, each member – be it non-exceptioned or exceptioned – shares its novel success with every other member of that class. This holds irrespective of whether the predicted phenomena were, on the one hand, known but not used in generating the theories or, even, on the other hand, unknown prior to being predicted. Finally, even in the latter case of predicting phenomena unobserved at the time, the claim that such temporally novel success affirms a theory's description of reality will be dramatically threatened in Section 5.5: there have been numerous theories in the history of science that are no longer taken to describe the world but achieved significant, even temporally novel, success: phlogiston theory, caloric theory, Dalton's atomic theory, Rankine's vortex theory, the massive-particle theory of light, to quickly note a few.

Were we to somehow find a justification for citing the mere descriptive fact that the non-exceptioned theories were those among the competitors that scientists happen to propose, we've only pushed the question to what the belief-relevant justification is for *proposing* empirically undistinguished non-exceptioned theories over their exceptioned competitors. We will now discuss a natural answer to this, one potentially looming beneath our discussion all along.

3.6 Virtue 6: Simplicity

It is hardly contentious that the demand for simplicity is a key supraempirical criterion that is both defied by the exceptioned competitors, and, in contrast, with others, a criterion employed in science. Nonetheless, claiming we can justifiably believe the simpler among competing theories, the scientific realist is faced with the burden of establishing that this preference with which we happen to find ourselves has anything at all to do with underlying reality. Here we arrive at that core issue: without imposing our mere preferences onto the world what grounds have we for such a demand?

There is widespread, if not universal, agreement among both realists and non-realists that claiming justified belief in simple over non-simple competitors takes the thesis that the world is simple to already be established. Van Fraassen, for instance, grants that simplicity "is obviously a criterion in theory choice, or at least a term in theory appraisal" (1980, p. 90). Without appeal to extra-scientific ideals, "certain metaphysical or theological views," he writes, "it is surely absurd to think that the world is more likely to be simple than complicated" (p. 90). He says, "the virtue, or patchwork of virtues, indicated by ["simplicity"] is a factor in theory appraisal, but does not indicate *special* features that make a theory more likely to be true (or empirically adequate)" [original italics] (p, 90). (See also Worrall (2000, p. 356) and Lipton (2004, p. 143).) While others may disagree with the last points here, they will generally grant the following: simplicity may be a pragmatic virtue, but to claim we can *justifiably believe* our simple descriptions of the world, over, say, their indefinitely many exceptioned competitors, is to assume that the world is simple. The question, of course, is what could possibly justify believing that it is.

In fact, the problem runs deeper than suggested thus far. Some might claim we can or even must begin with a metaphysical framework. Notably, however, even lifting a ban on such frameworks will not suffice: because metaphysical frameworks are themselves supraempirical and will possess – or be charged with not possessing – supraempirical virtues, the case for preferring a given framework over its competitors is bound to be built on precisely what that framework would be invoked to provide, for example, a justification for denying that exceptioned theories describe reality, taking as already established a solution to the problem in advance of offering one.

In regard to this problem of simplicity,[12] Paul Horwich attempts to switch the onus of proof onto the non-realist/instrumentalist. He says,

---

[12] Horwich is using "simplicity" in a very broad sense, to include simplicity, unity, coherence, and so on.

the mere fact (if it is a fact) that no one has (or could) come up with an argument for the evidential relevance of simplicity does not constitute a reason for doubting its relevance. The case for instrumentalism requires positive grounds for maintaining that simplicity is not an indicator of truth. (1991 pp. 11–12)

Note first that Horwich's proposal will bring us nowhere near a justification for scientific realism. Even if, somehow, that general thesis that the world is ultimately simple were acceptable, the realist must further justify the belief that the ontological domain to which a specific empirically undistinguished non-exceptioned theory pertains is a domain in which the world's non-exceptioned nature obtains, and in turn deny that it can be one genuinely described by an exceptioned competitor. So Horwich's proposal leaves us very far from scientific realism, that is, far from any justification for believing our best explanations.

More crucially while "onus of proof" arguments can be weak and inconclusive, Horwich's claim, I'm afraid, reduces to absurdity. We need only restate his point while replacing "simplicity" with any random, odd property a theory may have. And we can replace "instrumentalism" with "contemporary philosophy of science":

"the mere fact (if it is a fact) that no one has (or could) come up with an argument for the evidential relevance of" being in accordance with numerology "does not constitute a reason for doubting its relevance. The case for" contemporary philosophy of science "requires positive grounds for maintaining that" being in accordance with numerology "is not an indicator of truth."

We might replace simplicity with the property of being in accordance with astrology, the Bible, Gnosticism, or the study of tea leaves and crystal balls; the property of being conceived on earth, on blank rather than lined paper, between 6pm and 3am, or after eating vegetarian sushi rolls, and so on. Any bizarre property is a candidate. If I assert that such a property of a theory is required for it to describe reality, both you and Horwich will surely want to know my grounds for asserting this. If I tell you that, no, you need to provide evidence that the property is *not* relevant to our beliefs about reality, you will both laugh (or chase) me out of the room. Contra Horwich, the burden of proof regarding the issue of whether a property genuinely bears on reality is clearly on the party who makes the positive claim that some property does. Finally, we – the realists and instrumentalists alike – are by no means burdened to assert that a given bizarre property "is not an indicator of truth"; we need only ask on what grounds one could possibly justify believing it is.

Likewise, the non-realist does not, and need not, maintain that "simplicity is not an indicator of truth." With this, we recognize that part of van Fraassen's claim is an overstatement. The non-realist need not go so far as to *deny altogether* that the "features" referred to, when attributing simplicity to a theory, "make a theory more likely to be true." For doing so the non-realist invites the charge that she has invoked her own albeit vague and extremely minimal metaphysics. Apparently, Horwich is seizing on this slip, playing on van Fraassen's overstatement, effectively insisting that the two positions are on par, that just as with the realists, the non-realist must claim justified belief in an ultimately positive ontology about underlying reality. But the non-realist refrains from any substantial ontology about this realm. In the context of our concern, the non-realist need only press the epistemic question as to how we can possibly justify denying that exceptioned theories describe reality and believe their non-exceptioned, but empirically undistinguished, counterparts. Since non-realists need no underlying ontology for which they claim justified belief or that is sufficiently potent as to *deny* of any description-type that it can describe underlying reality, any hint at a non-realist need for even a weak underlying ontology, be it from van Fraassen or Horwich, is entirely superfluous. With one camp requiring an underlying ontology and other wholly refraining from one, the two positions are not even close to being on par. So the burden of proof cannot simply be pushed from one camp to the other. As illustrated with the bizarre properties just noted, the unfoundedness of the denial that exceptioned theories describe reality – a denial required of believing non-exceptioned theories – is key. If my efforts, provided thus far, to demonstrate the absurdity of switching the burden to the non-realist were to somehow remain insufficiently salient, the following discussion will make that absurdity especially clear.

In Section 3.1 I began evaluating a proposal for favoring belief in theories that possess various supraempirical virtues in general, over those that do not. We saw that Psillos explicitly advocates empirical testing to justify those virtues we happen to favor in our attempts to describe underlying reality: we look to see whether virtuous background theories "enjoyed" empirical "support" (Psillos, 1999, p. 172). We saw serious problems with that proposal, not least of which was that it was *merely a proposal*. That proposal however is particularly relevant to the problem of simplicity. That is, the realist is especially tempted to claim that we can *empirically* justify discarding the exceptioned competitors in the quest to describe reality. For instance, Musgrave suggests that so long as the principle of simplicity is sufficiently explicated – and here it would be clearly explicated as being non-exceptioned over exceptioned – it becomes "a metaphysical principle which can, at first remove so to speak, be empirically assessed: roughly speaking, it is acceptable metaphysics if theories constructed under its aegis are empirically successful, while theories which violate it are not" (1985, p. 203). Musgrave writes, "It may not be absurd to think that Nature is simple (in some carefully specified sense or senses), if we can point to the empirical success of science in vindication of our belief" (p. 204). However, our preliminary considerations, in Section 3.1, have bearing here, where we can now pursue them in greater depth. Unable to grant victory to scientific realists in advance, that is, unable to dogmatically assert that scientific realism has been established, the correlates of direct concern in the course of our empirical testing will be (a) the property of being non-exceptioned and (b) that of being empirically successful. One virtue in *this* proposal is that both (a) and (b) allow for correlatively precise positive and negative instances. It is a meta-hypothesis both of whose correlates are identifiable so such that we can empirically identify both confirming instances and disconfirming instances. The *primary meta-hypothesis* to be empirically tested is,

– *non-exceptioned theories are more often empirically successful than exceptioned theories.*

This looks promising, no doubt, especially given its virtue of genuine testability. Nonetheless, first, even if it could be empirically *established*, this meta-hypothesis would fail to do what realists need it to do; it would do nothing to imply that exceptioned theories cannot properly describe reality. Second – as indicated in Section 3.1 regarding virtues in general – allowing that the meta-hypothesis could be empirically established, that fact can stand unproblematically as a sole basis for non-realists to advocate selecting non-exceptioned theories, that basis being the empirical success to which that property, according to the meta-hypothesis, leads. As a very significant point then, but even treating it as merely preliminary: even if the meta-hypothesis were fully established, nothing whatsoever would be gained in favor of scientific realism per se. These two points in themselves stand as very serious problems for the realist's proposal: it simply would not do what the realist needs it to do.

Most importantly, however – and surprisingly I expect – *that meta-hypothesis is simply false.*

On this crucial point, note first, what we have seen from the start of this discussion regarding exceptioned theories, Section 2.4: it is the corpus of contemporary science itself that entails indefinitely many exceptioned yet altogether successful and fully accepted assertions. Again, as noted there, in extracting exceptioned theories from that corpus, qualifying no less than relations asserted to obtain *in enormously many instances with just a few exceptions* are those components of accepted science specifying relations that obtain *in only a single instance and never otherwise*: and, again, added to these two extremes are all the situations asserted to obtain in between. Hence, there are indefinitely many successful exceptioned theories *within accepted science*, each of which has greater success than its non-exceptioned counterpart.

Scientific Realism   39

Second, given the expression we extracted from that foundational point, and our further application of that expression (in which our favored non-exceptioned theories instantiate T), it is clear that there are *indefinitely many successful exceptioned theories* that are *external to* accepted science, each of which shares the empirical success of our favored non-exceptioned theories. That is the second point we secured in Section 2.4 and with which we have been running up to now: our non-exceptioned theories are empirically undistinguished from *indefinitely many* exceptioned competitors. The latter must be included in any tallying of empirically successful theories. And they will be radically greater in number, to an indefinitely enormous extent, than the number of successful non-exceptioned theories. The primary meta-hypothesis we are considering – *non-exceptioned theories are more often empirically successful than exceptioned theories* – that meta-hypothesis we have now shown to be clearly false.

Third, and finally, we can be sure that there are indefinitely many *non-exceptioned* theories not included in the corpus of science that stand as utter failures empirically, which can be generated at will and are taken to be blatantly false, say, "all massive objects repel all other massive objects," "all light is made up of repelling massive particles," "all humans are omnipotent." Truly, in want of non-exceptioned theories that are epic failures empirically, the possibilities are endless. Hence, in addition to the *primary meta-hypothesis* indented previously, three other *subsidiary meta-hypotheses*, are clearly false:

– *non-exceptioned theories are empirically successful*
– *exceptioned theories are empirically unsuccessful*
– *non-exceptioned theories are more likely to describe the world than not*

And, of course, given the three key points we've just seen, and the falsity of the primary meta-hypothesis, we appear wholly unable to empirically support the conclusion the scientific realist needs, that

We can justifiably deny that exceptioned theories describe the world, in favor of believing our non-exceptioned theories

In fact, significantly, while the *primary and three subsidiary* realist's meta-hypotheses are rendered false by the observed data to which *they* pertain, the exceptioned theories we have been discussing throughout, themselves – that is, the competitors to our empirically undistinguished non-exceptioned theories – are patently *not falsified* by the observed data to which *they* pertain. Hence, crucially, the *primary and subsidiary* meta-hypotheses the realist needs, empirically testable though they are, patently *cannot* be invoked to justify the denial that the exceptioned competitors genuinely describe reality. The empirical data of concern here can do nothing whatsoever to block these exceptioned theories as genuine competitors to our empirically undistinguished yet favored non-exceptioned theories. These points very strongly suggest, surprisingly, I expect, that no such empirical victory for the realist is possible; the quest for this kind of empirical justification for blocking the exceptioned competitors looks utterly futile.

Here we've continued to address the question posed in our sixth challenge to scientific realism (Section 2.3): "What justification have we for believing that the supraempirical criteria (realists claim) scientists favor in theory-choice have anything at all to do with reality?" Specifically, we've explored here six theoretical "virtues" to see whether our realist can invoke them to exclude the indefinitely many exceptioned competitors to our favored non-exceptioned theories. We first considered a *general* proposal pertaining to background systems: if our favored T shares the virtues of empirically successful background systems, it inherits that support. After seeing the failure of this general proposal, we turned to a set of specific theoretical virtues: *coherence* with background knowledge; a kind of indirect, vicarious support; breadth of scope; novelty; and now simplicity. In each case, we found that these supraempirical virtues will not suffice to justifiably exclude the competitors. We find ourselves with no justification for believing that the supraempirical criteria (realists claim) scientists favor in *choosing the best explanation* have anything at all to do with reality. With no such justification, the core question of scientific realism looms large: how can we possibly be justified in believing our best explanations – for example, over their indefinitely many *empirically unexcluded exceptioned competitors*?

4 Truth and the Argument from the Bad Lot

4.1 The Realist's Explicit Appeal to Truth

As has been clear, for the realist, a primary target of inference to the best explanation in science is underlying reality itself, the reality that lies beneath natural phenomena. Claiming justification for believing our best explanations, realists claim that what they believe describes that reality – a reality whose entities and processes, very explicitly for the realist, do not depend on what anyone believes about them. So far, discussing scientific *realism*, I've endeavored to maintain phrasing that explicitly keeps *reality* front and center. And apart from a few quotations, we have largely managed to leave the term "truth" to the side. Doing so accords with efforts by realists who explicitly resist that term. Exemplifying that effort is Michael Levin's (1984) argument that, with respect to the natural phenomena on which our realists have been focusing, it is the theory that is explanatory, not its truth: "Truth ... has nothing to do with it" (1984, p. 124). Devitt (2010, 2013), who's basic argument we saw in Section 1.1, expresses sympathy with this view, as do others such as Michel Ghins (2002), along with those embracing a deflationary theory of truth. Beyond keeping reality at the fore, another virtue of avoiding "truth" thus far is that we have given no false impression that such parties are somehow protected from the challenges we have surveyed. Nonetheless, as we now move forward, the appeal to truth takes a central role; and it is important to introduce the fact, at least here, that most contemporary realists explicitly invoke it. They will emphasize that despite our restraint – or the restraint of philosophers such as those just mentioned – from employing the term, it's been implicit all along: to believe P is to believe that P is true. Making that explicit now, the realism with which we are now concerned states explicitly that we can justifiably believe something *about* our best explanations of natural phenomena, namely that they are true.

4.2 Challenge 7: The Argument from the Bad Lot

In Section 1.5, touching on the argument from the underdetermination of theories by data, we looked briefly to Mill discussing the luminiferous ether. He wrote,

> Most thinkers of any degree of sobriety allow, that an hypothesis of this kind is not to be received as probably true because it accounts for all the known phenomena, since this is a condition sometimes fulfilled tolerably well by two conflicting hypotheses ... while there are probably a thousand more which are equally possible, but which, for want of anything analogous in our experience, our minds are unfitted to conceive. (Mill, 1867, 296)

Similarly – and, like Mill, mindful of the ether – Pierre Duhem asks,

> Do two hypotheses in physics ever constitute ... a strict dilemma? Shall we ever dare to assert that no other hypothesis is imaginable? Light may be a swarm of projectiles, or it may be a vibratory motion whose waves are propagated in a medium; is it forbidden to be anything else at all?

Duhem writes, "the physicist is never sure he [*sic*] has exhausted all the imaginable assumptions". (1906 [1954, p. 190])

In Section 1.5, along with our first look at Mill's comment, we also introduced the competitor thesis that challenges realism:

> *For a given T there are genuine competitors*, distinct alternatives whose descriptions of reality we cannot justifiably deny.

Stating it now in terms of truth makes clear just how natural our current shift is:

> *For a given T there are genuine competitors*, distinct alternatives whose truth we cannot justifiably deny.

Zeroing in on the challenge raised by Mill, Duhem, and the competitor thesis, now in the context of inference *to the truth* of the best explanation, we can see them effectively asking the following: what justification could we possibly have for believing the truth is included in the set of theories we are considering, which is, at the same time, to deny that it is instead in Mill's class of "probably a thousand more" "conflicting" but successful hypotheses "our minds are unfitted to conceive"; in Duhem's class of non-"exhausted" but "imaginable" hypotheses; or, jumping now to the end of the twentieth century and the start of the contemporary debate, in Lawrence Sklar's class of "unborn hypotheses" – the "innumerable alternatives to our best present theories" that "would save the data equally well" (1981, pp. 18–19).[13] Here we move toward an argument for which many of our previous considerations provide a foundation.

A particularly refined version of this argument has been expressed by van Fraassen, who has served as our primary foil against scientific realism and whose arguments I've used to springboard discussion. After introducing the realist's existence abductions, in Section 1.2 I set the stage with Challenge 1, van Fraassen's largely neglected argument against the demand for explanation. And in Section 2.2 with Challenge 5, against the realist's claim that we use IBE to justify belief in observables and so unobservables, I introduced van Fraassen's competing empirical hypothesis about our inferential practice. In Section 2.3 we also arrived at Challenge 6, which – using the exceptioned competitors identified in Section 2.4 as a tool – has since driven the bulk of our inquiry: how can we possibly justify imposing onto reality the supraempirical virtues we happen to find ourselves preferring. Each of these challenges traces back to van Fraassen's (1980), and our explorations regarding Challenge 6 pave the way for, and strongly connect with, another challenge to which we are now turning. This comes from van Fraassen's (1989) and accords in spirit with what we've just seen hinted at by Mill, Duhem, and Sklar. However, it drives home a key point only implicit in those. Though related to underdetermination, involving as it does a competitor thesis, this argument is treated as distinct – see, for instance, Psillos (1999), Lipton (2004), and Wray (2018), among others, such as Ladyman et al. (1994) – and, accordingly, we will treat it distinctly here, dubbing it Challenge 7.

Van Fraassen (1989) again challenges the realist's IBE as a rule for belief, and a primary argument there has come to be known by two names: *the argument*

---

[13] Stanford later (2006a, 2006b) employs the apt label "unconceived alternatives."

from the bad lot (e.g., Psillos, 1999) and *the argument from underconsideration* (Lipton, 1993/2004; Wray, 2018). Van Fraassen notes that, because IBE "only selects the best among the historically given hypotheses" (1989, p. 142–3), our theories cannot be put up against "those no one has proposed" (p. 143). That given, "our selection may well be" nothing more than "the best of a bad lot" (p. 143). Elaborating, he writes, "To believe" a hypothesis "is *at least* to consider" it to be "more likely to be true, than not" [original italics]. This means that "to believe the best explanation requires more than an evaluation of the given hypothesis," more than a "comparative judgement that this hypothesis is better than its actual rivals" – which, van Fraassen grants, "is indeed a 'weighing (in light of) the evidence'" (p. 143). We now arrive at van Fraassen's pivotal point: *believing* the best explanation "requires a step beyond" that comparative evaluation. In particular, it "requires a prior belief that the truth is," more likely than not, already included in the set of "actual," *available,* rivals (p. 143). He writes that, since this argument "is independent of the method of evaluation (of explanatoriness) that is used," any response will have to focus on that extra step, the *prior belief that the truth is already among the lot* of "historically given hypotheses" being compared – and that the truth is not, instead, among "those" rival hypotheses "no one has proposed" (p. 143), "never yet formulated" (p. 146) – that it is not instead among Sklar's "unborn hypotheses" (1981), Duhem's non- "exhausted" set of "imaginable assumptions," or even Mill's ""probably a thousand more" hypotheses "our minds are unfitted to conceive" (1867).

Psillos claims "the only reasonable interpretation of van Fraassen's argument" is that "*it is more likely that the truth lies in the space of hitherto unborn hypotheses*" [original italics] (1999, p. 217). However, in agreement with Ladyman et al. (1994), I submit that nothing in *this* argument of van Fraassen's even suggests such a conclusion.[14] Although van Fraassen is taking it to be a *genuine* (rather than merely logical) *possibility*, we must concur given our previous considerations, from Sections 2.3 to 3.6. Rather than any claim about what is likely, van Fraassen is claiming, as is clear in what we've just seen, that believing the best explanation "requires a step beyond" comparative evaluation, namely a "prior belief that the truth is," more likely than not, already included in the class of "the historically given" *available* and "actual" rivals, and not instead in the class of those "no one has proposed" (p. 143). What he is emphasizing is that *this extra but necessary step is left wholly unjustified.* Contra Psillos, this is no assertion about *what is likely*; it is rather a call to justify a step that scientific realists simply grant to themselves.

---

[14] Four pages later, van Fraassen provides an argument that does align with Psillos's take. However, Psillos himself recognizes it as distinct and addresses it separately (1999, p. 222).

A rather crucial interpretative question is what precisely van Fraassen is allowing – at least for the sake of argument – when saying "the comparative judgement" involves weighing the hypotheses "(in light of) the evidence" and noting his argument to be "independent of the method of evaluation (of explanatoriness) that is used" (p. 143). One might take him to be fully *allowing*, even if not granting, that supraempirical criteria qualify as grounds for saying one theory is more likely to be true than another. And realists have certainly read his argument this way. For instance, Lipton (1993/2004) surprisingly takes it as conceding to a "ranking premise," where the ranking is truth-relevant. However, I submit, the supposition that supraempirical virtues relate to truth does not follow from van Fraassen's words. One can allow, as van Fraassen does elsewhere, that various methods involving explanatory or pragmatic virtues may be employed, while nonetheless restricting "evidence" to empirical data: with regard to "the comparative judgement," some of the historically given hypotheses could be eliminated because they fail against the data; all the while "the method of evaluation" could involve supraempirical virtues. As the latter is a *descriptive* methodological point, and not about justified belief, it requires no concession or even allowance that those virtues are "evidential" or relevant to truth regarding unobservables. (See also Wray (2018).[15])

4.3 Dovetailing Challenges 2 and 7: History, Privilege, and the Bad Lot

In Section 1.3, we briefly glimpsed at the historical argument against realism, generally construed as a pessimistic meta-induction (PMI). There we saw comments by Mill (1859) and Tolstoy (1895). Summarizing such views, Henri Poincaré wrote, "the ephemeral nature of scientific theories takes by surprise the man [sic] of the world," who "sees them abandoned one after another ... ruins piled upon ruins," predicting "that the theories in fashion today will in a short time succumb" and "concludes that they are absolutely in vain. This is what he calls *the bankruptcy of science*" [original italics] (1902 [1907, p.160]). In Section 3.2, while discussing the realists background-in-place demand, we saw hints at Mill's and Tolstoy's prescience. There it was suggested – and I've argued elsewhere (2016b) – that Newton's theory was not merely appended but radically overthrown. Leaping forward from their anticipation of an overthrow, Kuhn explicitly dubs such radical changes "revolutions" (1962), and Hilary Putnam offers up an "overwhelmingly compelling" "meta-induction" (1976, p. 184). Mary Hesse turns this PMI into what

---

[15] In fact, I submit, my interpretation is wholly supported by van Fraassen's subsequent points in his text (1989, pp. 147–8).

Scientific Realism 45

she dubs *a principle of no privilege*: "our own scientific theories are held to be as much subject to radical conceptual change as past theories" (1976, p. 264). She writes, "this principle . . . arises from accepting the induction from the history of science" (p. 271). The point of course is that, as van Fraassen later puts it, we might "claim . . . privilege for our genius" and "glory in the belief that we are predisposed to hit on the right range of hypotheses" (1989, p. 143), to believe that finally, *now*, we've got the truth; however, says Hesse, "the revolutionary induction from the history of science about theory change" (1976, p. 268) pushes us to deny that privilege. The PMI in the present context can be expressed as follows:

1) We now see that our best past theories were false.
2) Therefore, we have reason to expect that our best contemporary theories are likewise false

I ended Section 2.2 with the suggestion that we keep watch for further opportunities for cumulative, dovetailing arguments. And we've seen a number of interrelations since – the competitor thesis, the bad lot argument, and arguments pertaining to simplicity, to name a few. Underdetermination, along with its close relative, van Fraassen's "underconsideration" or "bad lot" argument, have generally been treated as distinct from the PMI, with authors such as Psillos (1999) discussing them in entirely different chapters – and arguably, though in the same text, offering different versions of realism in response. While Mill had pointed to each, intriguingly Sklar, briefly mentioned in Section 4.2, combined the two in his, "Do Unborn Hypotheses Have Rights?" (1981), at the very start of contemporary realism debate. There, reflecting "upon historical scientific experience," he suggests "that there are innumerable alternatives to our best present theories" that "would save the data equally well" (pp. 18–19). Tying the pessimistic meta-induction into Sklar's alternatives that "save the data equally well," we find that "historical scientific experience" itself reveals a history of empirically undistinguished competitors. (I will build on this in Section 5.9.)

Given the PMI's premise, each past theory had competitors presumably "unborn" at the time. Hence the standard PMI entails Sklar's PMI regarding "unborn" competitors:

1) We now see that our best past theories had empirically undistinguished but "unborn" competitors.
2) Therefore, we have reason to expect that our best contemporary theories have empirically undistinguished but "unborn" competitors.

Now consider the implications for the bad lot argument posed by tying Sklar's induction into Hesse's no-privilege principle. In fact, the first realist "reaction" to the bad lot argument that van Fraassen considers – from which I borrowed his quip that bears repeating – is the realist "claim of privilege for our genius," the temptation to "to glory in the belief that we are by nature predisposed to hit on the right range of hypotheses" (1989, p. 143). Though van Fraassen happily avoids the PMI (2007, p. 347), dovetailing the strands here, we now have empirical evidence, a history of theories, that, despite being the best among the "lot" of "historically given" and "actual" explanations, were merely the best of a bad/false lot. Outside that lot was an explanation, *better* – at least in terms of empirical success combined with supraempirical/pragmatic virtues – than the *best* we had. It was instead in the lot of rivals "no one ha[d] proposed" (van Fraassen, 1989, p. 143), "never yet formulated" (p. 146), of "unborn hypotheses" (Sklar), of the non-"exhausted" set of "imaginable assumptions" (Duhem) – or even in the lot of those that scientists' "minds," were "unfitted to conceive" (Mill). (Stanford (2006a, 2006b) discusses "the problem of unconceived alternatives"; for my critique see (Lyons, 2014).) The "underconsideration problem," as Lipton dubs the problem of the bad lot (1993/2004, p. 153), permeates the history of science; replete with "unconsidered" competitors, it is a history of bad lots. We are led naturally to Hesse's "principle of no-privilege," now bolstered. Running the PMI with regard to such competitors, we would conclude that we have reason to expect that today's best theories are nothing more than the best of a bad lot.

4.4 A Shift toward Deductive Validity and a Clear Directing of the Burden

The historical arguments we've seen thus far – for instance, as supporting van Fraassen's challenge against realist privilege – are naturally challenged by realists pointing to the fact that they are inductive. However, I have argued elsewhere (Lyons, 2002, 2016a, 2017) that the historical argument should not be construed as a logically fallacious inductive inference to the falsity of our current theories. Instead, it is a (set of) logically valid *modus tollens* argument(s) that strikes directly at the thesis scientific realist's claim we can justifiably believe, along with the justification itself. While I will discuss this construal in Section 5.6, here I'll simply contend that we need neither make an inductive inference nor infer the falsity of our current best explanations or that they are, *in fact*, merely the best of a bad lot. Rather, in the present context, we see a serious historical threat to the realist claim to epistemic privilege and to the thesis that IBE reliably gives us unobservable truths. The realist hypothesis at this stage, that our best explanations are true, is challenged at nearly every turn in the history of science, be it by large-scale overthrows, or small-scale modifications occurring uncountably many times throughout history. Regarding the latter, consider the *normal* course of science as Kuhn (1962), followed by Lakatos (1970) – and I venture to say every philosopher of science now – construes it: although predictions derived from a theoretical system clash against data, the deep level theory goes unquestioned; the responsibility is on the scientists to "fit" the theory to the facts by modifying smaller scale hypotheses in the system to which the deep level theory is conjoined. Although Kuhn dubs this "puzzle solving," following Putnam's (1974) emphasis on the "explanatory" nature of the "schema" that Kuhn discerns, we can call that mode of reasoning "the explanatory schema." Despite other contentious elements in Kuhn's philosophy of science, there is little question from any camp that this explanatory schema captures the process involved in the *normal* course of science. Lipton's (1993/2004) thoughtful articulation of the nature of IBE, for instance, accords with it perfectly.

In Section 4.1 noting that to believe P is to believe that P is true, we made explicit the shift from the realist claim that we justifiably believe our best explanations to the claim that we justifiably believe those explanations are true. With that necessary clarification of the realist thesis, we now see that nearly every turn in the history of science fuels the *modus tollens* against that thesis for which our realist claims justified belief, and in turn against the claim to privilege and truth-reliability, including truths regarding the coveted realm of unobservables. And for present purposes anyway, taking our dovetailing effort to provide a non-inductive historical supplement to van Fraassen's argument, each of those turns in the history of science constitutes an instance in which the best that was inferred was, not just potentially, but *actually* merely the best of a bad lot. (While Wray relates the historical and underconsideration arguments, he embraces the pessimistic induction as such.) Along with resisting any induction and the unnecessary assertion that current theories are false, Hesse's *no*-privilege principle can be tempered. Privilege need not be outright *denied*; one need only note that, in claiming privilege, the burden is clearly on the realist to answer the challenges traced throughout this text, now including the history of inferring the best of a bad lot.

While contemporary scientific realists do not dovetail arguments in the manner I've done here, they do offer an answer to the challenge here posed, a pivotal meta-level argument to which we now turn.

5 The Realist Justification for Epistemic Privilege: The No-Miracles Argument

5.1 The Realist's Shift to the Meta-Level: Explaining *Success*

Have scientific realists any recourse by which they can ground a claim to epistemic privilege? Realists have taken steps to do just that. Mindful of some of the problems we have traced, realists are forced to concede that, even if scientists choose the best explanation, the lot from which it is selected may well be a bad one, with no explanation therein sufficing for an inference to truth. As Lipton aptly puts it, "the best" explanation must itself, for the realist, be "good enough" (2004, p. 56). The theories for which our realists now claim justified belief are not simply those accepted within science as the best available explanations. Rather they are those theories to which we can attribute a specific property, *empirical success*.

Appealing to *success* in need of justifying a claim to privilege, our realist now shifts to a *meta-level inference* we have not yet considered. While its significance is easily overlooked, this shift is ubiquitous in the contemporary literature. The realist is no longer claiming justified belief merely in theories that best explain *physical phenomena*. Nor is the realist explaining those natural phenomena by a property of those theories, *their truth*, and claiming, on those same explanatory grounds, justification for believing those theories are true. Rather that which is being explained has now shifted, one step removed from the level of natural phenomena, to a *property of theories* themselves: their *empirical success*. One property of theories, their *truth*, now explains another, their *success*. Shifting to this meta-level, our realist says it is not just the base-level IBE employed in science that justifies believing that the theory's unobservable entities exist; rather we look at the significant empirical success our theories have achieved and are struck by that success: *it would be a miracle were our theories to be as successful as they are were they not true*. Base-level IBE is replaced with this meta-level No-Miracles Argument (NMA).

It is here at this meta-level, explaining a property of theories, their *empirical success*, rather than the natural phenomena themselves, that, *by way of the NMA*, we are justified in believing theories have another property, truth. The realist NMA gives us a *meta-level justification* for *believing a meta-hypothesis*, a hypothesis *about* our scientific theories, involving two *properties* of theories: those that have the property of empirical success also have the property of truth. Notably, with this shift, the realist has eradicated a major problem that has plagued much of our exploration thus far. At the base-level, favored theories are faced with indefinitely many competitors; at the meta-level, however, our realist's meta-explanation, as expressed in the NMA, has no competitors whatsoever. According to the NMA, the truth of our theories provides *not merely the best* explanation for their success; rather – barring miracles, which no one in the debate accepts – it provides the *only* explanation. Moreover, we can now see the meta-level NMA as very explicitly addressing the realist burden of grounding the claim that we *do* have epistemic privilege. It is the NMA, pivoting on empirical success, that justifies our belief that those descriptions of underlying reality provided by our empirically successful theories are true

5.2 Challenge 8: Competing Explanations at the Meta-level: The Selectionist Explanation

Van Fraassen (1980) properly recognizes NMA as a distinct (meta-level) argument, dubbing it the "Ultimate Argument" for scientific realism (p. 39), yet he challenges it as well. Van Fraassen construes the question the NMA purports to answer as that of why the theories we have are successful. And his charge is that, endeavoring to answer this question, the realist has given us, not the only explanation, but only a false dichotomy. Contrary to the realist, we are not forced to choose between miracles and getting at truth about unobservables, with the latter being the only acceptable option. We have a third option, another explanation for success, a Darwinian selectionist explanation: "the success of scientific theories is no miracle. It is not even surprising to the scientific (Darwinist) mind. For any scientific theory is born into a life of fierce competition, a jungle red in tooth and claw. Only the successful theories survive..." (1980, p. 40). Why are the theories we have empirically successful? Because empirical success is a criterion for selecting them: we reject those that fail to be empirically successful. This third option between miracles and scientific realism is meant to wholly defuse the no-miracles argument. Because scientific realism is not providing the only explanation for the success of science, rejecting scientific realism does not leave us opting for miracles. Despite the fact that, as we have been tracing it, scientific realism has now shifted to this meta-level to explain a new phenomenon, empirical success, the realist explanation now and nonetheless has its own competitor, the selectionist explanation.

5.3 The Realist's Two-Pronged Defense: Truth as the Best Explanation for the Novel Success of Individual Theories

Realists respond, naturally, by claiming that this competitor is not as good as their own and that ultimately, even if realism does not provide the only explanation for success, it provides the best. Two specific realist responses to van Fraassen's selectionist explanation have emerged as prominent, both of which are offered by Alan Musgrave (1985).

The first is that the Darwinian explanation fails to explain *the success of an individual theory*. This point has been embraced by, not just Musgrave (1985, 1988), but also Lipton (1993/2004, 1994), and Leplin (1997).[16] Musgrave's second point in response: Even if theories have been selected for their empirical success, that selectionist explanation is unable to explain a particular kind of empirical success, namely *novel predictive success*.

We briefly discussed novel success in the context of competitors at the base-level of scientific theories. However, the realist has now raised the bar to this meta-level of, now, not merely general empirical success but a very specific kind, more difficult to obtain: novel predictive success. This – along with the claim that it is the success of specific theories that calls for explanation – serves not only to render the selectionist explanation insufficient, it is also well motivated: novel predictive success offers a genuine sense of wonder, and in the face of van Fraassen's alternative explanation, genuinely bolsters the NMA. In fact, it also appears to block the anti-realist's historical argument, giving substance to the claim that theories must be good enough: they must achieve novel predictive success. This virtue was not lost on Musgrave. Writing against Laudan who in 1981 offered a now infamous list – which we briefly referenced in Section 1.3 – of successful theories that are patently false, Musgrave writes, "few arguably none, of the theories cited" by Laudan "had any *novel* predictive success" [original italics] (1985, p. 211, ftnt 10). Addressing all these problems in one fell swoop, Musgrave likewise brings genuine potency to the realist effort to ground epistemic privilege. *It would be a miracle were our theories to achieve novel predictive success were they not true.*

Although Wray (2018) attempts to revitalize van Fraassen's explanation as answering these demands, he separates them from one another. However, Musgrave's challenge involves their conjunction: what calls for explanation is *the novel success of an individual theory*. I don't see that Wray succeeds. In fact, in terms of novel success, when van Fraassen later challenges the realist's claim to privilege, implicitly referencing his selectionist alternative, he says, the "jungle red in tooth and claw does not select for internal virtues – not even ones that could increase the chance of adaptation or even survival beyond the short run" (1989, p. 143). Likewise, Wray himself writes that in "the biological world," on the basis of a species past success "we are apt to be surprised if it does not continue to survive, *unless we are aware of changes in the*

---

[16] In fact, van Fraassen concedes to it immediately upon introducing it (1980, 40, ftnt 34)

environment ..." [my italics] (2018, p. 169). Such changes in environment, however, are precisely what temporally novel predictive successes are meant to constitute. And at one point Wray concludes that the novel success of false theories would simply be "mysterious" (p. 170). Our realist would agree, deeming this a euphemism for "miraculous."

5.4 Reprise, Challenge 3: Truth, Degree of Implication, and the Problem of Explanatory Vacuity

In Section 1.4, considering Peircean abduction, in which a posit renders phenomena a matter of course, that is, at least likely, we glimpsed at what I dubbed Challenge 3. The issue there was an explanatory posit's *degree of implication* for phenomena – the degree to which that which does the explaining implies that which is explained. Mindful that this would bear on our realist's explanatory demand, we noted that an existential posit of an unobservable would have to also include, at the very minimum, a set of property descriptions, which we dubbed "a theory." In Section 2.1, adding the insight that theory evaluations are comparative, we shifted to IBE at that base-level of scientific inference. In that context, and bringing these points together, one can expect that, assuming other explanatory virtues are equal, a theory possessing a greater *degree of implication* for a set of phenomena than another stands as the better explanation. We return to that issue here, but now shifted up to this meta-level, where the phenomenon calling for explanation is *novel predictive success of particular theories*. As noted, this shift alleviates us from facing the indefinitely many competitors at the base-level. However, we've now seen a competitor at this meta-level and the realist charge that it cannot explain the relevant phenomena. Tying these points together, I suggest that this realist charge is effectively that the selectionist explanation has an extremely low degree of implication for the novel success of a particular theory; rejecting theories that have not enjoyed empirical success fails to do anything to render likely the novel predictive success of a theory that has. Further, and by contrast, we can take the realist to be saying that "T is true," does not merely have a high degree of implication for novel success, it may even entail it. Statements implying the obviousness of entailment are offered on both sides. Laudan, in his critique of realism, takes it to be "self evident": "if a theory is true, then it will be successful" (1981, p. 30). Musgrave claims it is "obvious that a true theory will be successful – after all, true premises yield true conclusions" (2017, p. 91). Perhaps, most commonly, it is taken to be so obvious as to go without saying.

Challenge 3, however, prompts us to inquire about the degree to which "T is true" implies novel predictive success. While in Section 1.4 we settled for "theory" as consisting of, at least, property attributions conjoined to existential statements, we would want, say, Newton's law of universal gravitation conjoined to his three laws of motion, or Einstein's field equations, to qualify as "core theories." Nonetheless, these core theories alone assert nothing at all about, say, the way in which a point of light in the vast sky will behave. As Duhem emphasized, to derive a range of even basic explanations and predictions one will often need to conjoin to a theory a full theoretical edifice, vastly many other theoretical statements: other universal auxiliary statements, initial conditions, idealizations, and so on. And this is especially so for a deep theory about unobservables. Yet the stipulation that core theory, T, "is true" by itself fails to specify anything about any such auxiliaries, so strictly speaking it fails to *require any*; with no such restrictions, a true theory need not lead to any empirical predictions. Further, even if we take for granted that auxiliaries are conjoined to T, "T is true" does nothing to restrict the *content* of those auxiliaries; so even with auxiliaries it need not engender empirical predictions. Beyond that, even explicitly adding that the auxiliaries permit T to bring about empirical predictions, that addition does nothing to require that those predictions can be, let alone have been, tested in ways humans can implement. Moreover, since "T is true," does nothing to preclude as auxiliaries patently false statements, even if we allow auxiliaries such that testable empirical predictions can be derived, that mere fact does nothing to entail that they are confirmed. In fact, needing to consider all possible combinations of all possible auxiliaries that could in principle be conjoined to T to render empirical consequences, there are indefinitely many false auxiliaries and combinations that would render T *unsuccessful*, irrespective of T's truth. And since those auxiliaries and combinations that would make empirical success likely are limited, the number of possible auxiliaries and combinations that would engender predictive success is radically lower than the number that would bring failure; it is vastly more likely that a true core theory would fail than be successful. Finally, even if the edifice conjoined to our true theory brought about some true predictions, nothing guarantees that our observation statements accord with them, without which confirmation would not obtain. Adding to all this, the demand for novelty only decreases the degree of implication "T is true" has, especially, for temporally novel success. We simply cannot say that "T is true" makes empirical success, let alone such a strict form of success, even the least bit likely. Contrary to Laudan, the anti-realist, and Musgrave, the realist, it is not at all "self-evident" or "obvious" that the mere stipulation "T is true," entails even general predictive success, let alone novel success. And that is so even adding the numerous further stipulations we have generously granted regarding auxiliaries.

5.5 Reprise, Challenge 2: Novel Success from False Theories

Let us momentarily bracket the issue of a theory's degree of implication for success, as discussed in the previous section. Musgrave's appeal to novel success was no doubt a game changer for the realist, posing a serious challenge to non-realists. As we saw, it threatens, not only van Fraassen's selectionist explanation, Challenge 8 in Section 5.2, but also the historical argument against realism, Challenge 2 in Sections 1.3 and 4.3. In fact, although Musgrave introduced novel success to the debate in the mid-1980's, a major deficit in the arguments put forward by most antirealists, to this day, is that they fail to attend to novel success, at least when it counts; and neglecting that, they fail to answer this realist meta-hypothesis. Among them are van Fraassen, as well as more contemporary antirealists such as Kyle Stanford (2006a, 2006b) (see my (2006b, ftnt 10) and (2014)) and Wray (2018). Most importantly, as Musgrave emphasizes, Laudan was not concerned with novel success, and he is the one who, following Hesse and Putnam, most significantly stressed the historical argument (1981). All the while, most realists have followed Musgrave's call to focus on novel success.

Since this realist focus is so well-motivated, for instance, increasing the potency of the realist's no-miracles argument (NMA) and apparently eliminating items on the historical list, I've spent much of my own work taking up Musgrave's challenge that "few arguably none, of the theories cited" by Laudan "had any *novel* predictive success" [original italics] (1985, p. 211, ftnt 10). In fact, in a series of publications I've detailed various novel predictive successes that were derived from theories which by present lights can only be taken to be false. Included among them are, for instance, phlogiston theory, caloric theory, Dalton's atomic theory, Kekulé's theory of the benzene molecule, Mendeleev's periodic law, Fresnel's wave theory of light and theory of the optical ether, Fermat's principle of least time, Bohr's 1913 theory of the atom, the original (pre-inflationary) big bang theory, W.J.M. Rankine's vortex theory, Dirac's relativistic wave equation and hole theory (for each of these see Lyons (2001), (2002)); Kepler's theory of the *anima motrix*, Newtonian mechanics, Adams and Leverrier's solutions to the problem of the behavior of Uranus (Lyons (2006b)); Pasteur's theory that where there is optical asymmetry there is life; Maxwell's mechanical theory of the ether regarding vortices and idle wheels, Descartes conception of a God who created extension and instilled motion into the world, Scheele's specific version of Phlogiston theory, Lamark's theory that catastrophes and mass extinctions have not occurred in Earth's history, Thales's posit that water is "the arche" conjoined to auxiliaries pertaining to, for instance, divinity and seminal principles, the Miller–Urey theory of the early earth (Lyons (2014, ftnt 13)); the Newtonian theory of light as massive pellets that vary in speed, as dependent on their mass; the Schwarzschild solution to Einstein's field equation; Einstein's cosmological solution, a static, homogeneously filled, spherical universe; and de Sitter's cosmological solution, a hyperbolic, and taken "literally," "completely empty universe" (Lyons (2016b), (2017)). Each of these theories is false but each led to significant novel predictive successes. In fact, restricting my list here to the most rigidly demanding sort, *temporally* novel success, such false theories have led to some of the greatest successes in the history of science. (See also the collection of newly uncovered examples in (Lyons and Vickers, 2021).) It should be clear that the far less demanding use-novelty, to which realists also appeal, extends the list (potentially much) further.

Allow me to briefly circle back to Section 5.4, where I argued "T is true" has a low degree of implication for novel success. We now see the futility of a natural realist response, that what realists obviously mean when they say "T is true" is "T and all its auxiliaries are true." This assertion is wholly untenable, even more strongly refuted by such examples and innumerably many beyond those listed here. Noting this, the tension here, to which we will return, is crucial: whether because of an explanatory vacuity due to an extremely low degree of implication or because of a failure to explain these successes due to the falsity of the theories, the realist claim, *'T is true' explains and is needed to explain novel predictive success* fails. Although it appears that the selectionist has a low degree of implication for the novel success of individual theories, the selectionist can at least explain why we have theories that have been successful. By contrast, now in light of the tension we are seeing for the realist, it is unclear whether the realist can explain even that. I offer this here as a significant supplement to van Fraassen's and Wray's selectionist explanation.

5.6 Reprise Challenge 2, the Historical Meta-*Modus Tollens*

Noting the historical threat to scientific realism, we have seen it expressed as a pessimistic meta-induction (PMI) in Sections 1.3 and 4.3. And most authors – though they will occasionally shift around on this – construe the realist argument as just that: an induction from past falsity to the falsity of contemporary theories. However, as noted in Section 4.4, I've argued that this is a poor construal of the argument. With the NMA now in hand, I can articulate how I contend the argument should be construed: it is a set of deductively valid meta-*modus tollens* arguments that challenge the meta-hypothesis the realist claims we can justifiably believe, along with the realist's justification for believing it, the NMA. The latter, in the form we are now considering, is, *it would be*

Scientific Realism 55

a miracle were our theories to achieve novel predictive success were they false. One version (introduced in my 2001, 2002) is best seen as a bi-layered (2016a, 2017) meta-*modus tollens*; the two layers are separated in the following argument by semi-colons in the premises and conclusion. The first layer pertains directly to the realist's meta-hypothesis, while the second pertains to the no-miracles argument:

1) If (a) the realist meta-hypothesis holds, then (b) we would not find instances of false theories achieving novel predictive success; for, given the NMA, those would be "miracles."
2) However, (not-b) we do find instances of false theories achieving novel predictive success; we have a list, which given the NMA is a list of "miracles."
3) Therefore, (not-a) the realist meta-hypothesis is false; and, barring miracles as we all are, the realist justification for believing it, the NMA, is rendered unacceptable.

Given the unacceptability of the NMA, it does not justify believing the realist meta-hypothesis; and because the realist meta-hypothesis is false, we cannot justifiably believe it – irrespective of the NMA. The meta-hypothesis fails to survive even as a mere postulate or defeasible conjecture. Psillos (2016) and Psillos and Ruttkamp-Bloem (2017) – mindful of my efforts to uncover increasingly many historical cases that threaten various realist meta-hypotheses – challenge my proposal that the historical argument is a meta-*modus tollens*: "it makes the past record of science irrelevant" (2017, p. 3191). The idea is that, if it is deductively valid, we can have no evidential gradation against realism of the kind afforded by a pessimistic meta-induction; so my effort to bring forward numerous past cases is pointless. However, even bracketing here two additional versions of the meta-*modus tollens* offered in my (2016a, 2017), the importance of increasing the historical instances holds even for the original (2001, 2002). Not only does such an increase secure the second premise, the bi-layered nature of the argument makes explicit that the NMA is no less a primary target: we have a list of "miracles," increasingly many novel successes that are, for the realist, inexplicable, despite the realists' heavily advertised insistence that only they can explain them; and barring miracles as we all are, but given the realist's demand for explanation, some non-realist explanation is required for each false theory achieving novel success. Further, since that non-realist explanation will have to be able to explain successes of theories past *and* present, its promise for explanatory breadth is vastly (and increasingly) greater than that of the realist explanation, which, again, wholly fails to explain each instance in the list; the greater the quantity, the greater the failure for the realist explanation. All this holds despite the deductive validity of the *modus tollens*. And since the key claim of the no-miracles argument is false, it is patently unacceptable as providing justification for believing the (likewise false) realists' meta-hypothesis. To these points I've added (2017, 2018) that each counterinstance diffuses the intuitiveness of the NMA, increasingly eliminating the residual psychological hope for it, to which the realists cling. I will revisit and further the historical argument in the next few sections.

5.7 Reprise, Challenges 2 and 3: Approximate Truth, Degree of Implication, and the Historical Argument

Since the meta-hypothesis against which I'm currently invoking these examples is "theories that achieve novel success are true," the realist has a natural move in their arsenal, invoking approximate truth rather than truth per se. Musgrave himself resists this: he expects that false theories have not led to novel success and considers approximate truth to be problematically vague. However, other realists invoke it regularly – at least when they're engaging with Laudan, who tends to explicitly discuss approximate truth, less so when they're engaging with van Fraassen, who, we have seen, tends to discuss truth per se. In any case, the aforementioned lists, coupled with the fact that today's very best theories cannot be true per se, *forces* the realist to appeal to approximate truth. On the latter note, our very best theories, general relativity and quantum field theory, are in significant conflict with one another (see my 2016a), and at least one is at best only approximately true. Yet I doubt it can be denied that they are the most empirically successful theories humankind has ever proposed.

However, I contend, even the appeal to approximate truth does not suffice. I've argued in the relevant texts that most if not all of my examples of theories achieving novel success are rendered by present lights such that they cannot even be construed as approximately true; they are *patently false*.

Moreover, and importantly, Laudan did not connect his critique of the downward path, pertaining to what we have discussed as a theory's degree of implication, to his critique of the upward path, the historical argument, which he also but far more famously discussed in his (1981). Otherwise untapped then is the fact that the argument regarding the low degree of implication can be leveraged against the historical meta-*modus tollens*, whose list is one of successes that are inexplicable for realism. In Section 5.4, I argued that the stipulation "T is true," unexpectedly, does not have a high degree of implication for novel success, that, with only that stipulation, a deep level theory is more likely to fail than be successful – no doubt a surprising conclusion to those including Laudan and Musgrave who take truth to fully entail success. There our primary concern was with auxiliaries, about which any stipulation that "core theory, T, is true" says nothing. Nonetheless, I took this further and noted that the threat remains even granting numerous additional stipulations regarding auxiliaries. Turning now to approximate truth, every point in Section 5.4 holds no less, and attending now to the core theory itself and the stipulation that *T approximates the truth*, the situation becomes considerably worse. To illustrate, grant the wholly implausible assumption that every auxiliary to which our core theory is connected is true per se. A core theory that is approximately true, *even when it is connected only to true auxiliaries*, need not be successful.

Here is an illustration: Replicate the corpus of successful contemporary science indefinitely many times. Separate that collection of new theoretical systems into four subsets. In each system in the first subset, *raise* the charge posited for the electron. In one of those system's assign a value one one-thousandth higher than in the original corpus; and in another assign it a value one one-billionth higher than the original. For each of the indefinitely many remaining systems in this first subset, assign a distinct value for the electron charge that falls between these values. In the systems of the second subset, *lower* the electron charge the same way. In the third and fourth subsets, do as with the first two except change the charge posited for the *proton*. Our full collection of previously replicated systems now consists of indefinitely many that approximate the original corpus. However, each theoretical system so closely approximating our contemporary corpus predicts that matter repels matter, so no universe whatsoever or at least one wholly unlike ours – constituting dramatic empirical failure. Even keeping all auxiliaries and background theories identical, the slightest change in claims about unobservables need by no means lead to empirical approximation. Allowing next that the many auxiliaries only approximate their relative postulates in the original corpus, the situation is compounded. Finally, adding the points we've seen in Section 5.4, the situation is clearly worse than ever for the realist. Given these additional problems, the stipulation that a core theory is approximately true is considerably less likely to imply its success than even the stipulation that it is true.

Further, all this is so even while requiring that, to be approximately true, a theory must "refer." Relinquishing that requirement inflates by multitudes the quantity of theories and systems that qualify as approximately true. The degree of implication that "T (or System) is approximately true" has for novel success would be diminished even more radically, rendering the claim that approximate truth explains novel success wholly untenable. This appears to put a restriction on our notion of approximate truth: an approximately true theory must refer – and do so in a substantial sense. Although as I mentioned Laudan discussed this problem in different terms in his (1981), challenging the realist's downward path, his point is that realists have not shown that the approximate truth of T would entail its success. My conclusion here and in Section 5.4 is much stronger: *we have good reason to deny* that the mere stipulation that T is approximately true (or even true) will render success even (merely) likely. And although Laudan simply takes for granted that approximate truth requires reference, I've now given grounds for accepting this.

While this is an important point, returning now to the historical argument, my list of theories that achieved novel success but are false, many patently so, does not require denying reference. Irrespective of reference, the appeal to approximate truth does not suffice: most if not all of my examples of theories achieving novel success cannot, by present lights, be construed as even approximately true. As noted, Laudan left his critiques of the downward and upward paths unconnected. I've argued however that the argument regarding the low degree of implication that the realist postulates have for success can be leveraged against the historical meta-*modus tollens*, whose list is one of successes that are inexplicable for realism: Combining the arguments, we see that diluting approximation, increasing its permissiveness, to accommodate the list makes approximate truth so vacuous as to fail to render success likely, destroying its degree of implication. By contrast, narrowing approximate truth so that it renders success likely eliminates its touted explanatory breadth, increasing miracles. Either way, whether because of explanatory vacuity or items on the list fueling the meta-*modus tollens*, the realist is unable to explain success, so unable to offer the best, let alone the only, explanation (see Lyons, 2003, 2016b, 2018). Despite the move to approximate truth, then, we have numerous theories whose novel success realism fails to explain.

5.8 Reprise, Challenge 2: A Historical and Evidentially Graded Meta-*Modus Ponens*

In light of the refutation of the realist's meta-hypothesis by way of the meta-*modus tollens*, realists will be tempted to insert within their original meta-hypothesis the qualification "statistically likely": the meta-hypothesis we can justifiably believe is merely that "our theories achieving novel predictive success are statistically like to be approximately true." This immunizing tactic appears to allow the realist to ignore any item on the historical list and the original meta-*modus tollens*. Even then there are further meta-*modus tollens* that strike at a realism embracing this immunized meta-hypothesis (Lyons, 2017). In want of answering both the move to statistical likelihood and Psillos's denial that my logically valid historical argument affords no evidential

Scientific Realism 59

increase, here I introduce a historical meta-*modus ponens* that accounts for *evidential gradation*.

First, mindful of the comparative insight discussed in Section 2.1, and mindful that non-realists refrain from commitment to the existence or non-existence of specific unobservables, take now what I will call an *outrageous anti-realist guess*, to which no anti-realist should commit: "all of our theories achieving novel predictive success are patently false." We can deploy that guess purely for the sake of comparative evaluation. Second, as indicated when considering theoretical virtues in Sections 3.1 and 3.6, without simply granting victory to realism in advance of the debate, we can identify no correlatively precise confirming instances of "approximate truth" about unobservables that correlate with supraempirical virtues. Of current concern here is "novel predictive success" of which we *can* empirically identify positive instances (though, as noted in Section 3.5, it is not supraempirical). And we can secure the correlatively precise "patent falsity" in light of the fact that our past and present theories both achieve novel success but radically contradict one another; *doing so requires no specification of where the falsity lies*. Hence, the list provides correlatively precise positive instances of the outrageous guess as the competitor. And those items stand also as correlatively precise negative instances for the realist's statistical meta-hypothesis. Moreover, again without granting victory to realism from the start, we have no correlatively precise negative instances of that outrageous guess. With this I offer a logically valid meta-*modus ponens* that strikes at scientific realism itself – in this case, the claim that we can justifiably believe the realist's meta-hypothesis that statistically correlates approximate truth and novel success – while at the same time reinforcing the importance of increasing the items on the list. *The historical and evidentially gradated* meta-*modus ponens:*

1) If (a) we have greater – or much, vastly, or overwhelmingly greater – evidence for the outrageous anti-realist guess, "all of our theories achieving novel predictive success are patently false," then (b), clearly, we cannot justifiably believe our realist meta-hypothesis, namely, "our theories achieving novel success are statistically likely to be approximately true."

2) However, (a) we do have greater evidence – or, *as we increase the items on the list*, much, or vastly, or overwhelmingly greater evidence – for that outrageous anti-realist guess.

3) Therefore, (b) we cannot justifiably believe the realist meta-hypothesis, in this case that "our theories achieving novel success are statistically likely to be approximately true."

Again, the outrageous guess is simply used as a tool for comparative evaluation. And not only does this meta-*modus ponens* refute in a logically valid manner the claim that we are justified in believing the realist's statistical meta-hypothesis, increasing the items on the list moves us toward the increasingly threatening options in its premises, from greater to overwhelmingly greater evidence. We have no logically fallacious induction but a logically valid meta-*modus ponens*, and we have reason to increase *the evidential gradation*: each case *increases the extent to which our lack of justification* for the realist meta-hypothesis *is evident*.

As a final note, this argument makes no mention of the NMA. However, not only does the immunizing statistical realist meta-hypothesis threaten to defy the realist insistence that their position is empirically testable (in any way other than I've just tested it), because it permits inexhaustibly many "miracles," it wholly sacrifices the sole justification realist have for believing their meta-hypothesis, the NMA. If this new realist statistical meta-hypothesis survives the *original* meta-*modus tollens*, it does so at a radical and unacceptable cost: the realist is left with no justification for believing it. Epistemic scientific realism has been sacrificed.

5.9 Reprise, Dovetailing Challenges 2 and 7: History and a Bad Lot *Modus Ponens*

We concluded Section 4 by dovetailing a set of non-realist arguments. Specifically, we combined Hesse's historical induction with van Fraassen's argument from the bad lot or underconsideration, which explicitly pertains to competitors; doing so, we articulated a Sklar-like historical induction regarding competitors that denies the claim to epistemic privilege scientific realism requires. Against that challenge, however, in Section 5.3, we began considering a refined realist meta-hypothesis and NMA, in which it is specifically novel success that, barring miracles, contemporary realists claim a patently false theory could not achieve. We are now prompted to recognize base-level competitors in the context of the aforementioned historical arguments and the list of, what are by present lights, patently false theories achieving novel success. In this new context, we can consider a second noteworthy set of *competitors*, this time not extracted purely from within the corpus of contemporary science *per se*, but from *the relation(s) between contemporary and past scientific theories*.

Although any among those past theories achieving novel success might suffice, my favored example for addressing this issue (Lyons, 2014, 2016a) is Kepler's deep theory of the *anima motrix*. In the course of Kepler's reasoning, he centrally

Scientific Realism 61

included the following posits, each of which is patently false by contemporary lights: the sun is a divine being and the center of the universe; the natural state of the planets is rest; there is a non-attractive emanation from the sun, the *anima motrix*, that pushes the planets forward in their paths; the planets are inclined to be at rest, so to resist the solar push, and this inclination contributes to their slowing speed when more distant from the sun; the *anima motrix* pushing the planets is a "directive" non-attractive (and he later adds) magnetic force, and so on. Among the novel predictions Kepler makes are that the sun spins; it spins in the direction of planetary motion; it spins along the plane of the ecliptic; and it spins faster than any of the planets revolve around it. Beyond those novel predictions, his deep theory was also centrally deployed toward his laws, themselves crucial to the unprecedented success of the Rudolphine Tables. Those laws led to, and continue to lead to, innumerable successful predictions pertaining to the behavior of, not only Mars and Earth, but also Mercury, Venus, Saturn, and Jupiter. Early on, Kepler achieved further significant novel successes – pertaining to relations between the Earth, Sun, and planets – predicting, not only two planetary transits, the Mercurial, and the rare and irregular Venusian, transit, but also a separation between the two transits of less than a month. And his laws led, and continue to lead, to numerous successful novel predictions regarding the then undiscovered planets, Uranus and Neptune, as well as any number of additional bodies in the solar system and beyond. (For more detail, see Lyons (2006b).)

Consider now the way in which the status of a past theory such as Kepler's is to be expressed from the context of contemporary science. While contemporary theory, CT, patently contradicts the content of Kepler's theory, T, it shares but goes beyond the successful predictions of Kepler's theory. The following is expressed by CT:

The phenomena are (approximately) as T predicts, except in situations S, in which case the phenomena behave in manner M.

This expression, extracted from the standpoint of a contemporary theory, CT, captures a *relation between past and present theories* that will hold for each theory in our list of patently false but nonetheless successful historical theories. With Kepler's theory instantiating T, for instance, we can include as S occasions in which, say, Jupiter approaches Saturn and in which a planet's orbit is particularly close to the sun, and so on; and we can add, as M, "non-Keplerian perturbations," "the advancement of Mercury's perihelion," and so on. Articulating S and M in *full* detail, these will be *complicated assertions* expressed by CT. However, reinforcing our considerations in Sections 3.1 to 3.6, it is clear that, *unless realists deny the approximate truth of CT*, realists must concede that the absence of supraempirical virtues such as simplicity affords no grounds to deny the approximate truth of these very complicated assertions.

Stepping now from our contemporary context back to a historical perspective, and employing the expression just extracted, we have to concede that it asserts a genuine competitor to Kepler's theory, T. Contemporary science itself reveals a competitor, CT, which, though contradicting T, shares those predictions successfully made by T. Moreover, it is clear that, beyond the specifics provided by our contemporary corpus, there are also indefinitely many alternative S's and M's available that are entirely out of accord with contemporary science. Nonetheless, from our historical vantage point, they too are competitors to Kepler's theory.

Returning now to our contemporary perspective, we realize that, just as the previous expression reveals competitors to such past theories, it also reveals such competitors to any contemporary theory, CT. Instantiating T now with any accepted CT, the remaining clauses can include any S that has not (yet) been acknowledged as obtaining and any M that significantly differs from the behavior our favored CT describes. There are indefinitely many options and combinations, all of which will share our favored CT's predictions about observed phenomena. This process we have traced, pertaining to the relation between present and past successful theories reveals that there are indefinitely many competitors to any contemporary theory we may favor.

It is clear that this method for revealing competitors is *historically informed*, using as it does the relation between present theories and the list of past theories achieving novel success that are, by contemporary lights, rendered patently false. And again, one can say a past theory is *false by present lights* without making any commitment to the (approximate) truth of present theories. One need only recognize that we have competitors that patently contradict one another, fully allowing that neither is even approximately true. Though the relations between them are discerned by syntactic analysis, they are historically exemplified relations, obtaining between an empirically identified past scientific theory and an empirically identified contemporary theory. This method also and nevertheless remains *non-inductive*: the historically exemplified, empirically identified competitor relation extends by way of instantiation to those contemporary theories that are related to phenomena in the way that scientific theories are non-contentiously required to relate to phenomena – no induction required. In fact, that non-inductive commitment can be retained while extending our effort in Sections 4.3 and 4.4 to dovetail the historical and bad lot arguments. That is, our empirically but non-inductively revealed competitors can be integrated into the following logically valid argument,[17] a *bad lot modus ponens*:

1) Past theory T qualifies as a candidate for the realist meta-hypothesis, that is, the type of theory realists claim we can justifiably believe is approximately true (e.g., past T has enjoyed novel success).
2) If, however, we have reason to believe that past T has genuine competitors, those whose approximate truth we cannot justifiably deny, then *we have no grounds to deny* that T is merely the best of a bad – patently false – lot, leaving us with no justification for believing T is approximately true.
3) For past theory T (which qualifies as a candidate for the realist meta-hypothesis), contemporary theory, CT, expresses the following competitor: *the phenomena are (approximately) as T predicts, except in situations, S, in which case the phenomena behave in manner, M.*
4) Contemporary theory, CT, can be instantiated as T in that expression (italicized in (3)), affording indefinitely many variants, no induction involved.
5) We have (every) reason to believe that contemporary theory, CT, has indefinitely many genuine competitors, those whose approximate truth we cannot justifiably deny.
6) Therefore, we have no grounds to deny that CT is only the best of a bad – patently false – lot; we are left with no justification for believing contemporary theory, CT, is approximately true.

With no inductions, by way of the historically exemplified, empirically identified relation between contemporary and past successful scientific theories, this bad lot argument nonetheless retains a genuine connection to the historical argument. Despite the realists' move to novel predictive success of specific theories and to approximate truth, the realist claim to epistemic privilege is lost.

5.10 Reprise, Challenge 8: Modest Surrealism and a Supplementary Competitor Explanation for Success

In Section 5.2 we noted that once we have a competitor for the realist explanation of success, we have shifted from the NMA to a meta-level IBE. Like nearly all realists, van Fraassen and Wray do not discuss an explanation's degree of implication. However, the points in Sections 5.4 and 5.7, as noted, provide a partial response on behalf of the selectionist meta-explanation: although its

---

[17] I introduced an argument along these lines in my (2014), where I engage with and critique Stanford's (2006a, 2006b) version of a Sklar-like but double-inductive bad lot argument that emphasizes the inability of particular scientists to conceive of alternatives. As I point out, however, Stanford neglects showing the novel predictive success of the theories – such as Kepler's in my example – *to which* his unconceived alternatives are alternatives.

degree of implication for the novel success of individual theories may be very low, the realist explanations appear to fare no better, or even worse.

Barring miracles, just how would we explain the novel success of patently false individual theories? The non-realist explanation I favor is a modest version of one offered by Arthur Fine (1986), introduced in my (Lyons, 2002, 2003, 2018), which – using Leplin's (1987) tongue-in-cheek phrase for Fine's – I call *modest surrealism* (MS):

> *the mechanisms postulated by the theoretical system would, if actual, bring about – or have among their consequences – the relevant phenomena observed, and some that will, but have yet to be, observed at time t; and these phenomena are brought about by actual mechanisms in the world*

A system would simply not qualify for MS were its auxiliaries to render its predictions of observed phenomena unconfirmed. Provided those phenomena are sufficiently wide-ranging, a theoretical system with the property captured in MS ("TS is MS") will have achieved both the general and novel predictive success at a given time; and, compared to its meta-competitors, will have a far higher degree of implication for phenomena that will be but are not yet observed: "TS is MS" implies novel success to a far greater degree than "T is true," "T is approximately true," as well as "T was selected for (past) success." (I put MS up against other meta-competitors in my (2003).) Crucially, since MS withstands the list of historical successes rendered inexplicable by the realist meta-explanations considered, MS explains each such success. In fact, I suggest MS is the natural explanation for the success of a theory that is patently false, and for that matter, *any* theory, irrespective of whether it *might be* true – wholly eliminating any need to invoke (approximate) truth regarding specifics about unobservable reality.

One more point: the base-level competitor thesis (discussed previously) also explains the novel success of patently false historical theories: the latter are among the indefinitely many mutually non-approximating theories whose specific range of successes are shared by our contemporary theories (Lyons (2016a)). Extending this, given the second set of competitors in which we instantiated current theories for T, we can say contemporary theories are successful because they are among the set of indefinitely many theories that would be successful. In fact, I've suggested (Lyons 2018) that one can take this as, essentially, a syntactic expression of the modest surrealist's semantic meta-explanation. We now have two related meta-explanations, one semantic and one syntactic. Put up against the realist's appeal to the approximate truth of the theory, both have a higher degree of implication for novel success; and, explaining the novel success of potentially true *and* false theories, both have far greater explanatory breadth.

6 Conclusion and Epilogue: Socratic Scientific Realism

6.1 Conclusion

Since scientific realism is the default view of most philosophers involved in the realism debate, and is thought, at least by realists, to be the default view of scientists, we've concerned ourselves heavily with challenges to scientific realism, tracing along the way a set of realist sophistications in response to those challenges. These sophistications in turn prompted a set of novel variations on, and novel instances of dovetailing, our collection of eight non-realist challenges. Throughout the course of our inquiry, we've seen a set of what appear to be solid blows against each of the realist sophistications we've explored. Consequently, and in conclusion, I suggest that none of these variants of scientific realism can be considered tenable, including even the realist's most dramatic and immunizing retreat to the meta-hypothesis, "our scientific theories achieving novel success are statistically likely to be at least approximately true."

Accepting this conclusion, we ask, where can scientific realism go from here? Two promising candidates are deployment realism, introduced by Kitcher (1993) and Psillos (1999), and epistemic structural realism introduced by Worrall (1989). Although I critique the former in my (2006b), (2009), and (2017) and the latter in my (2016b), it may surprise the reader that, despite all of the challenges I've raised thus far, I nonetheless consider myself to be a scientific realist.

6.2 Epilogue: Socratic Scientific Realism

Taking the epistemic tenet of scientific realism to be seriously threatened, I endeavor to bracket our 2,500-year-old obsession with justifying belief about reality. Instead, I advocate, as a tool for inquiry, a purely axiological realism (2001, 2005), what I now call *Socratic scientific realism* (2016a, 2017, 2018, 2019). In Section 1.1, I began with the statement, "The central claim of scientific realism is that science endeavors to accurately describe reality beyond the realm of what we have observed or even can observe." That axiological meta-hypothesis is central to Socratic scientific realism. However, according to my meta-hypothesis, the end toward which scientific reasoning is directed is not truth *per se*. Instead, it is a particular sub-class of true statements, those whose *truth is experientially concretized*, abbreviated as *XT-statements*. These are true statements, including those about unobservables,

> whose truth is made to deductively impact, is deductively pushed to and enters into, *documented reports of specific experiences*, 'DRSEs'.

Supplementing the familiar notion of truth preservation, the deductive "pushing" here constitutes theorizing: replacing, modifying, adding low- or high-level auxiliary hypotheses or even core theoretical components. And the deductive "entering into" is the forging of logical connections by way of mediating terms. In the course of theorizing, the truth of a statement S is *made* to deductively impact a DRSE.

In addition to being true, and so excluding false statements, XT-statements can neither be *vacuous* nor altogether *detached* from a theoretical system. Moreover, crucially, they cannot be such that their truth fails to deductively reach any DRSEs *due to obstruction by false statements* in the theoretical system. The *extent* to which an XT-statement is experientially concretized can be understood as the *range of DRSEs* on which its truth is made to have deductive impact. Beyond that, there are varying *degrees* to which an XT-statement can be made experientially concretized, where the "degree of experiential concretization" references the gradation to which an XT-statement has specification toward, and is impacting on, DRSEs.

Importantly, no claim is made that we can discern just which statements in a theoretical system are in fact XT-statements. For instance, it is not the case that a statement's impact on DRSEs *informs us* of the statement's truth. (This is emphatically not an epistemic realism.) Nonetheless, one can sometimes discern when and roughly where we have a *deficiency* of XT-statements. With one type of evident XT-deficiencies, it is evident that non-XT statements are present in the theory complex. With another type, it is evident that we possess DRSEs that have no matching prediction statements.

In slightly more detail then, my axiological realist postulate is that science, in the course of modifying its theoretical systems, endeavors to remedy such evident XT-deficiencies by increasing the number – and/or the extent, degree, or exactitude of the experiential concretization – of XT statements; to retain or increase the extent and degree of the experiential concretization of each individual XT statement; to retain unaffected non-vacuous and non-detached statements; and to avoid increasing the non-XT (and the non-concretization of XT) statements. A modification of a theoretical system actually achieving these conditions constitutes an *increase in experientially concretized truth*, an IncXT. In short then, my postulate is that system modifications in science are *directed toward* achieving that state, an IncXT, be it at the deepest or most surface level of a theoretical system.[18]

Since this postulate, articulated in more detail in my (2005, 2011, 2019), is meant to be a specification of the end toward which scientific inquiry is directed,

---

[18] Fully embraced here is the recognition that theoretical choice can often be a gradual process, made after extended articulation and comparison.

its content is not meant to be particularly surprising. Rather, its novelty lies in the fact that achieving that goal requires and promotes certain theoretical virtues whose relation to one another and especially to truth are otherwise unclear. I've shown elsewhere (2005, 2011, 2019) that a set of eleven syntactic desiderata are required or at least promoted by the endeavor to achieve this end. More specifically, I show that the actual achievement of an IncXT entails and hence *requires* the achievement of an increase in empirical accuracy and consistency, and an increase in, or at least the retention of, breadth of scope, testability, and three forms of simplicity. I also show that the quest for an IncXT *promotes*, but does not require, a fourth form of simplicity, temporally novel predictions, explanatory depth, and an increase in, what I've called in previous sections, a system's degree of implication toward DRSEs.

To at least indicate the relation between actually achieving an IncXT and such virtues, consider first, even if briefly, an *increase in empirical accuracy:* taking that to refer to the number, breadth, and precision of prediction statements that match DRSEs,[19] and given that the truth of a statement cannot be experientially concretized where no such match obtains, an increase in empirical accuracy must obtain given the achievement of an IncXT. With that noted, here we can focus on the requirement of *an increase or at least retention* of one form of *simplicity,* specifically the form that would exclude the exceptioned competitors we've seen threaten realists throughout (competitors introduced in Section 2.4, as well as those in Section 5.9). While a complex containing our favored non-exceptioned theory would share its empirical success with indefinitely many complexes, each of which is identical except for the fact that it contains its own exceptioned variation of that theory, the switch from the former complex to any of the latter would be prohibited in pursuit of an IncXT: even if a given exceptioned theory and, hence, its exception clause were true, connected to our complex, and not such that its truth fails to reach empirical predictions due to obstruction by false statements, that exception clause cannot be such that its truth is experientially concretized, that is, made to deductively impact DRSEs. Thus, accepting the exception clause and, hence, the exceptioned theory, we would be accepting a theory that could not constitute an IncXT. While it remains possible that our favored non-exceptioned theory is false, if it is true, connected to a complex, and not such that its truth fails to reach DRSEs, for instance, due to obstruction by false statements, then in contrast with the exception clause, its truth is also experientially concretized. The portion of our non-exceptioned theory that goes beyond what has been experientially concretized at a given time is still part of

---

[19] Of course, such matches will often occur only via "bridges," for example, auxiliary statements regarding the margin of error. Also, since *empirical accuracy* is a syntactic relationship that does not pertain to all observables, it is distinct from van Fraassen's *empirical adequacy*.

a statement whose truth has (many) instances of experiential concretization, that is, whose truth has been made to deductively impact (many) DRSEs; no part of the exception clause has such instances. Thus, rejecting the indefinitely many empirically undistinguished exceptioned competitors as replacements for our non-exceptioned theories is *required* of the quest for an IncXT.

Stepping further, note that, even if the endeavor toward a particular end does not require a specific virtue but nonetheless *encourages* it, that endeavor would also provide good reasons for avoiding systems that defy that virtue. Accepting this, it is not only that, in the absence of distinguishing data, the pursuit of an IncXT gives us a reason for rejecting exceptioned theories as replacements; it also provides a reason for not proposing exceptioned theories as alternatives in the first place, irrespective of whether they might be tested. Taking theory proposal to be directed toward the primary goal, we would want to consider *just how well*, to what extent, a theory would meet that goal, just in case it does. And our goal dictates as significant the *extent* to which the truth of each statement in a complex is experientially concretized. Comparing statements, we ask what this measure *would be* for each statement if that statement were true and if its truth were experientially concretized in a set of DRSEs that distinguish it from its competitors. We recognize that, under these conditions, the instances of truth concretization will be divided between the individual components of an exceptioned theory, and the extent of concretization for each component would be lower than it would be for our non-exceptioned theory. The components of the exceptioned theory, if true and experientially concretized in distinguishing DRSEs, can never be concretized to the same extent that our non-exceptioned theory can be, if the latter were true and experientially concretized as such in distinguishing DRSEs. Thus, the endeavor to achieve an IncXT provides a reason to refrain from proposing exceptioned theories as alternatives to, or instead of, non-exceptioned theories, despite a lack of distinguishing data. Our posit then has implications not only for acceptance in terms of theory or system modifications, but also for the practices tacitly employed in *proposing* modifications to our theoretical systems.

Now our goal also, and quite appropriately, informs us of just when we are required to accept an exception clause: when we have a set of distinguishing DRSEs in which an exception clause is such that, if true, its truth is experientially concretized. Even in such an instance, we see now that our goal will push us to seek a new statement which is such that, if true, its truth is experientially concretized to a greater *extent* than would be the potential truth of the exception clause and to a greater *extent* than would be the potential truth of the statement to which the latter marks off an exception. Hence, when data does favor exceptioned over non-exceptioned theories, the quest for an IncXT provides reasons

Scientific Realism 69

for seeking deeper-level non-exceptioned theories that encompass those exceptioned theories.

Seeking an IncXT, we also have reason to exclude large-scale systems of exceptioned theories that would be empirically undistinguished from our own. First, given no distinguishing DRSEs, even if such a system and so its many exception clauses were true, those exception clauses are not such that their truth is experientially concretized, that is, made to deductively impact DRSEs. Thus, accepting the exceptioned system, we could not be accepting a system that constitutes an IncXT. Second, as earlier, we ask what the extent of concretization *would be* for each statement in a complex were that statement true and were its truth experientially concretized in a set of DRSEs that distinguish it from its competitors. And the individual statements of such an exceptioned competitor system, if true and experientially concretized as such in distinguishing DRSEs, cannot themselves be concretized to as great an extent as can those individual statements in our theory complex, if true and experientially concretized in distinguishing DRSEs. The quest for an IncXT gives a reason then to exclude and refrain from proposing such exceptioned competitor systems.

In sum then, while the standard realist postulate that science seeks truth wholly fails to account for these practices, the postulate that science seeks an IncXT offers such an account. Hence it offers a *purely axiological solution* to *the problem of simplicity*, minimally, as we've seen here, the kind of simplicity of concern throughout much of our inquiry: The *endeavor to achieve an IncXT* provides good reasons to

- refrain from replacing our non-exceptioned theories with their indefinitely many empirically undistinguished exceptioned competitors (in fact, the quest for an IncXT *requires* this);
- refrain from proposing the many possible empirically undistinguished exceptioned theories as alternatives in the first place;
- propose deeper-level non-exceptioned theories that encompass those exceptioned theories the DRSEs do favor; and
- exclude and refrain from proposing full systems of exceptioned theories that are empirically undistinguished from our own system.

Because my refined realist meta-hypothesis requires or promotes a full collection of eleven virtues – including those briefly discussed here, empirical accuracy and a crucial form of simplicity – it offers a unifying end, one that provides good reasons, and hence justifies and accounts for, not only the individual factors in theory choice, but their collection as well. If a proposed complex modification fails to meet, for instance, the specified necessary conditions for attaining an IncXT, that modification defies the posited end toward which scientific practice is directed. Going well beyond the ubiquitous practices just noted, I contend that my axiological meta-hypothesis both justifies and accounts for that larger collection of eleven virtues.[20] Consequently, I contend it amounts to advocating a non-belief-based – and, once those virtues are unpacked, a far more informative articulation of – IBE (2006a, 2012) which, along with the scientific corpus IBE produces, constitutes a tool for inquiry, that is, a tool for *increasing experientially concretized truth*. In fact, I contend that this axiological meta-hypothesis better accounts for and justifies the processes in science than, not only the belief-based realism we've considered throughout, but also non-realist axiologies offered by van Fraassen (1980), Laudan (1996), and Hoyningen-Huene (2014). The battle cry for Socratic scientific realism is that science seeks truth without claiming to possess it; its pursuit is justified, irrespective of whether we can justifiably believe we have achieved it. These last components take considerable argumentative work – see, for instance, my (2005), (2017), (2019) – which takes us beyond the scope of the current text. Nonetheless, I contend that there is a variant of scientific realism that wholly withstands the serious challenges we've surveyed throughout: Socratic scientific realism.

---

[20] Included in this explanatory package is the hypothesis that the relevant practices are taken to be required of "the type of truth science seeks" – my axiological postulate being an articulation of that type of truth. For an empirical defense of this corresponding hypothesis see my (2001, 2005).

References

Boyd, R. (1973). Realism, underdetermination and the causal theory of evidence. *Nous*, **7**, 1–12.

Cartwright, N. (1983). *How the Laws of Physics Lie*. New York: Oxford University Press.

(1999). *The Dappled World*. Cambridge: Cambridge University Press.

Devitt, M. (2010). *Putting Metaphysics First: Essays on Metaphysics and Epistemology*. Oxford: Oxford University Press.

(2013). Realism/anti-realism. In M. Curd and S. Psillos, eds., 2nd ed., *The Routledge Companion to Philosophy of Science*. London: Routledge, pp. 256–67.

Duhem, P. (1906 [1954]). *The Aim and Structure of Physical Theory*. P. Wiener (trans.), Princeton: Princeton University Press.

Feyerabend (1963). How to be a good empiricist: A plea for tolerance in matters epistemological. In B. Baumrin, ed., *Philosophy of Science: The Delaware Seminar, Volume 2*, New York: Interscience Press, pp. 3–39.

Fine, A. (1986). Unnatural attitudes: Realist and instrumentalist attachments to science. *Mind*, **95**, 149–79.

Ghins, M. (2002). Putnam's no-miracle argument: A critique. In S. Clarke and T. D. Lyons, eds., *Recent Themes in the Philosophy of Science: Scientific Realism and Commonsense*. Dordrecht: Springer, pp. 121–37.

Glymour, C. (1984). Explanation and realism. In J. Leplin, ed., *Scientific Realism*. Berkeley: California University Press, pp. 173–92.

Haufe, C. (2016). Testing structural realism. In C. Haufe, ed., *Special Section, Studies in History and Philosophy of Science Part A*, **59**.

Hesse, M. (1976). Truth and the growth of scientific knowledge. *PSA: Proceedings of the Biennial Meeting of the Philosophy of Science Association*, **2**, 261–80.

Horwich, P. (1991). On the nature and norms of theoretical commitment. *Philosophy of Science*, **58**, 1–14.

Hoyningen-Huene, P. (2014). *Systematicity: The Nature of Science*. Oxford: Oxford University Press.

Khalifa, K. (2010). Default privilege and bad lots: Underconsideration and explanatory inference. *International Studies in the Philosophy of Science*, **24**, 91–105.

Kitcher, P. (1993). *The Advancement of Science*. Oxford: Oxford University Press.

References

Kuhn, T. S. (1962). *The Structure of Scientific Revolutions*. Chicago: University of Chicago Press.

— (1974). Logic of discovery or psychology of research? In P. A. Schilpp, ed., *The Philosophy of Karl Popper, The Library of Living Philosophers*, Vol. 14, Book 2. La Salle: Open Court, pp. 798–819.

Ladyman, J., Douven, I., Horsten, L., and van Fraassen, B. (1997). A defense of van Fraassen's critique of abductive reasoning: Reply to psillos. *The Philosophical Quarterly*, **47**, 305–21.

Lakatos, I. (1970). Falsification and the methodology of scientific research programmes. In I. Lakatos and A. Musgrave, eds., *Criticism and the Growth of Knowledge*. Cambridge: Cambridge University Press, pp. 91–195.

— (1974). Popper on demarcation and induction. In P. A. Schilpp, ed., *The Philosophy of Karl Popper, The Library of Living Philosophers*, Vol. 14, Book 1. La Salle: Open Court, pp. 241–73.

Laudan, L. (1981). A confutation of convergent realism. *Philosophy of Science*, **48**, 19–41.

— (1996). *Beyond Positivism and Relativism: Theory, Method, and Evidence*. Boulder: Westview Press.

— (2004). The epistemic, the cognitive, and the social. In P. Machamer and G. Wolters, eds., *Science, Values, and Objectivity*. Pittsburgh: Pittsburgh University Press, pp. 14–23.

Laudan, L. and Leplin, J. (1991). Empirical Equivalence and Underdetermination. *Journal of Philosophy*, **88**, 449–72.

Leplin, J. (1987). Surrealism. *Mind*, **96**, 519–24.

— (1997). *A Novel Defense of Scientific Realism*. Oxford: Oxford University Press.

Levin, M. (1984). What kind of explanation is truth? In J. Leplin, ed., *Scientific Realism*. Berkeley: California University Press, pp. 124–39.

Lipton, P. (1993/2004). *Inference to the Best Explanation*. London: Routledge.

— (1994). Truth, existence, and the best explanation. In A. A. Derkson, ed., *The Scientific Realism of Rom Harré*. Tilburg: Tilburg University Press, pp. 89–111.

Lyons, T. D. (2001). *The Epistemological and Axiological Tenets of Scientific Realism*, Ph.D. Thesis, University of Melbourne, Australia.

— (2002). Scientific realism and the pessimistic meta-*modus tollens*. In S. Clarke and T. Lyons, eds., *Recent Themes in the Philosophy of Science: Scientific Realism and Commonsense*. Dordrecht: Springer, pp. 63–90.

(2003). Explaining the success of a scientific theory. *Philosophy of Science*, **70**(5), 891–901.

(2005). Toward a purely axiological scientific realism. *Erkenntnis*, **63**, 167–204.

(2006a). Review Peter Lipton's *Inference to the Best Explanation*. *The British Journal for the Philosophy of Science*, **57**(1), 255–8.

(2006b). Scientific realism and the *Stratagema de Divide et Impera*. *The British Journal for the Philosophy of Science*, **57**(3), 537–60.

(2009). Non-competitor conditions in the scientific realism debate. *International Studies in the Philosophy of Science*, **23**(1), 65–84.

(2012). Axiological scientific realism and methodological prescription. In H. W. de Regt, ed., *EPSA Philosophy of Science: Amsterdam 2009*. Dordrecht: Springer, pp. 187–97.

(2014). The historically informed *modus ponens* against scientific realism: Articulation, critique, and restoration. *International Studies in Philosophy of Science*, **27**(4), 369–92.

(2016a). Scientific realism. In P. Humphries, ed., *Oxford Handbook of Philosophy of Science*. New York: Oxford University Press, pp. 564–84.

(2016b). Structural realism versus deployment realism: A comparative evaluation. *Studies in History and Philosophy of Science Part A*, **59**, 95–105.

(2017). Selectivity, historical threats, and the non-epistemic tenets of scientific realism. *Synthese*, **194**, 3203–19.

(2018). Four challenges to scientific realism and the Socratic alternative. *Spontaneous Generations: A Journal for the History and Philosophy of Science*, **9**(1), 146–50.

(2019). Systematicity theory meets Socratic scientific realism: The systematic quest for truth. *Synthese*, **196**, 833–61.

Lyons, T. D. and Vickers, P. (2021). *Contemporary Scientific Realism: The Challenge from the History of Science*. New York: Oxford University Press.

Maxwell, J. C. (1873). *Treatise on Electricity and Magnetism*, Vol. 2, London: Macmillan.

(1890). *Collected Scientific Papers of James Clarke Maxwell*. New York: Dover.

Maxwell, N. (1999). Has science established that the universe is comprehensible? *Cogito*, **13**(2), 139–45.

McMullin, E. (1984). A case for scientific realism. In J. Leplin, ed., *Scientific Realism*. Berkeley: California University Press, pp. 8–40.

(1991). Comment: Selective anti-realism. *Philosophical Studies*, **61**, 97–108.

Mill, J. S. (1859 [1998]). *On Liberty.* New York: Oxford University Press.

(1867). *A System of Logic.* New York: Harper.

Mullis, K. B. (1993). Press release. NobelPrize.org. www.nobelprize.org/prizes/chemistry/1993/summary/.

Musgrave, A. (1985). Realism versus constructive empiricism. In P. Churchland and C. Hooker, eds., *Images of Science.* Chicago: Chicago University Press, pp. 197–221.

(1988). The ultimate argument. In R. Nola, ed., *Relativism and Realism in Science.* Dordrecht: Kluwer, pp. 229–52.

(2017). Strict empiricism versus explanation in science. In E. Agazzi, ed., *Varieties of Scientific Realism: Objectivity and Truth in Science.* Switzerland: Springer, pp. 71–93.

Nichols, D. E. (2016). Psychedelics. *Pharmacological Reviews,* **68**(2), 264–355.

Peirce, C. S. (1958). *Collected Papers. Vol. 5.* Cambridge, MA: Harvard University Press.

Poincaré, H. (1902 [1907]). *Science and Hypothesis.* New York: The Walter Scott.

Popper, K. (1959). *The Logic of Scientific Discovery.* New York: Basic Books.

Psillos, S. (1999). *Scientific Realism: How Science Tracks Truth.* London: Routledge.

(2016). From the evidence of history to the history of evidence: Re-thinking the pessimistic X-duction. *Presented Feb 19 2016 at The History of Science and Contemporary Scientific Realism Conference,* Indiana University-Purdue University Indianapolis.

(2018). Realism and theory change in science, *Stanford Encyclopedia of Philosophy.*

Psillos, S. and Ruttkamp-Bloem, E. (2017). Scientific realism: quo vadis? Introduction: New thinking about scientific realism. *Synthese,* **194**, 3187–201.

Putnam, H. (1974). The corroboration of theories. In P. A. Schilpp, ed., *The Library of Living Philosophers, Vol. XIV, The Philosophy of Karl Popper.* LaSalle: Open Court, pp. 221–40.

(1976). What is "realism?" *Proceedings of the Aristotelian Society,* **76**(1), 177–94.

Reichenbach, H. (1930). Kausalität und Wahrscheinlichkeit *Erkenntnis,* **1**, 158–88.

Salmon, W. (1984). *Scientific Explanation and the Causal Structure of the World.* Princeton: Princeton University Press.

(1985). Empiricism: The key question. In N. Rescher, ed., *The Heritage of Logical Positivisms, Lanham: University Press of America,* pp. 1–21.

Sellars, W. (1962). *Science, Perception and Reality.* Atascadaro: Ridgeview.

References

Sklar. L. (1981). Do unborn hypotheses have rights? *Pacific Philosophical Quarterly*, **62**, 17–29.

Smart, J. J. C. (1963). *Philosophy and Scientific Realism*. London: Routledge & Kegan Paul.

(1968). *Between Science and Philosophy*. New York: Random House.

(1979). Difficulties for realism in the philosophy of science. *Logic, Methodology and Philosophy of Science VI*, **104**, 363–75.

Stanford, K. (2006a). Darwin's pangenesis and the problem of unconceived alternatives. *The British Journal for the Philosophy of Science*, **57**, 121–44.

(2006b). *Exceeding Our Grasp*. Oxford: Oxford University Press.

Swinburne R. (1997). *Simplicity as Evidence of Truth*. Milwaukee: Marquette University Press.

(2001). *Epistemic Justification*. Oxford: Oxford University Press.

Thagard, P. (1992). *Conceptual Revolutions*. Princeton: Princeton University Press.

Tolstoy, L. (1893 [1905]). The non-acting. In L. Wiener, ed., *The Complete Works of Count Tolstoy, Volume 23, Miscellaneous Letters and Essays, Translated from the Original Russian and edited by Leo Wiener*. Boston: Dana Estes, pp. 43–65.

(1895 [1903]). The non-acting. *Essays & Letters*, trans. A. Maud. London: Grant Richards, pp. 94–122.

van Fraassen, B. (1980). *The Scientific Image*. Oxford: Oxford University Press.

(1985). Empiricism in philosophy of science. In P. Churchland and C. Hooker, eds., *Images of Science*. Chicago: University of Chicago Press, pp. 245–308.

(1989). *Laws and Symmetry*. Oxford: Oxford University Press.

(2007). From a view of science to a new empiricism. In B. Monton, ed., *Images of Empiricism*. Oxford: Oxford University Press, pp. 337–83.

Wray, K. B. (2018). *Resisting Scientific Realism*. Cambridge: Cambridge University Press.

Worrall, J. (1989). Structural realism: The best of both worlds? In D. Papineau, ed., *Philosophy of Science*. Oxford: Oxford University Press, pp. 139–65.

(2000). Pragmatic factors in theory-acceptance. In W. H. Newton-Smith, ed., *A Companion to the Philosophy of Science*. Oxford: Blackwell, pp. 349–57.

Acknowledgments

In memory of my parents, David and Carole Lyons. Dedicated to my sons, Huxley and Everson Lyons. This work was supported by the Stephen J. Kern Programmatic Fund for Philosophy. Very special thanks to Tanya and Madison Ignacek, as well as my RAs past and present, especially, Hannah Ray, Adam Hayden, Jared Storm, Michael Seidel, Amanda Galloway, and John Hanks. I am grateful to the two anonymous reviewers for their especially helpful feedback and insights.

Cambridge Elements ≡

Philosophy of Science

Jacob Stegenga
University of Cambridge
Jacob Stegenga is a Reader in the Department of History and Philosophy of Science at the University of Cambridge. He has published widely on fundamental topics in reasoning and rationality and philosophical problems in medicine and biology. Prior to joining Cambridge he taught in the United States and Canada, and he received his PhD from the University of California San Diego.

About the Series
This series of Elements in Philosophy of Science provides an extensive overview of the themes, topics and debates which constitute the philosophy of science. Distinguished specialists provide an up-to-date summary of the results of current research on their topics, as well as offering their own take on those topics and drawing original conclusions.

Cambridge Elements =

Philosophy of Science

Elements in the Series

Scientific Representation
James Nguyen and Roman Frigg

Philosophy of Open Science
Sabina Leonelli

Natural Kinds
Muhammad Ali Khalidi

Scientific Progress
Darrell P. Rowbottom

Modelling Scientific Communities
Cailin O'Connor

Logical Empiricism as Scientific Philosophy
Alan W. Richardson

Scientific Models and Decision Making
Eric Winsberg and Stephanie Harvard

Science and the Public
Angela Potochnik

Feminist Philosophy of Science
Anke Bueter

Abductive Reasoning in Science
Finnur Dellsén

The Social Dimensions of Scientific Knowledge: Consensus, Controversy, and Coproduction
Boaz Miller

Scientific Realism
Timothy D. Lyons

A full series listing is available at: www.cambridge.org/EPSC

Printed in the United States
by Baker & Taylor Publisher Services